MW01527710

Contributions within Density Functional Theory with Applications in Chemical Reactivity Theory and Electronegativity

by
Mihai V. Putz

ISBN: 1-58112- 184-9

DISSERTATION.COM

Parkland, FL • USA • 2003

Contributions within Density Functional Theory
with Applications in Chemical Reactivity Theory and Electronegativity

Copyright © 2003 Mihai V. Putz
All rights reserved.

This work was carefully produced. However, author and publisher do not warrant
the statements, data, illustration, or other items to be free of errors. Readers are advised to
keep in mind that information contained therein may inadvertently be inaccurate.

Dissertation.com
USA • 2003

ISBN: 1-58112- 184-9
www.Dissertation.com/library/1121849a.htm

Seria C Nr 0005719

ROMÂNIA
MINISTERUL EDUCAȚIEI ȘI CERCETĂRII
The Ministry of Education and Research / Ministère de l'Éducation et de la Recherche / Ministerium für Bildung und Forschung

DIPLOMA DE DOCTOR

UNIVERSITATEA
The University / L'Université / Die Universität

UNIVERSITATEA DE VEST DIN TIMISOARA

ca instituție organizatoare de doctorat,
as an institution authorised to organise doctoral programmes / en sa qualité d'établissement organisateur d'études doctorales/
als für das Promotionsverfahren berechtigte Hochschule

conferă titlul științific de **DOCTOR**
confers the academic title of Doctor / confère le grade de DOCTEUR / verleiht den akademischen Grad eines Doktors

în domeniul .... CHIMIE / CHEMISTRY
in the field of / en / im Bereich

cu toate onorurile și drepturile aferente,
with all the associated rights and privileges / avec tous les honneurs et les droits afferents /
mit allen daraus hervorgehenden Ehren und Rechten

PUTZ V. MIHAI VIOREL

upon Mrs., Ms. / à Mme, Mlle / an Frau
upon Mr./ à M. / an Herrn

născut(ă) la data de .... 10.03.1975 .... în .... ROMÂNIA
born on / né(e) le / geboren am   in / en / in   țara / country / pays / Land

urmare a susținerii tezei de doctorat
following the successful defence of his / her doctoral thesis / suite à la soutenance de la thèse de doctorat /
als Folge der mündlichen Verteidigung der Dissertation

și în baza *Ordinului Ministrului Educației și Cercetării* nr. .4198. din 29.07.2002.,
and on the basis of Order no. ..., dated .... of the Ministry of Education and Research /
et en application des dispositions de l'arrêté no. ... du ... du Ministre de l'Éducation et de la Recherche/
aufgrund der Verordnung des Ministers für Bildung und Forschung Nr. ... vom ...

Rector,
Président de l'Université / Rektor

Secretar șef,
Registrar / Secrétaire général / Chefsekretär

Nr. 107 din 05.02.2003

Limba oficială de pregătire a doctoratului, Conducătorul de doctorat, Titlul tezei de doctorat, Data susținerii tezei de doctorat.

Limba oficială de pregătire a doctoratului .... ROMÂNĂ
Official language of the doctoral programme / Langue officielle du programme / Offizielle Sprache des Promotionsstudiums

Conducătorul de doctorat .... Prof.dr..ing.. ADRIAN. CHIRIAC...........................
Supervisor / Directeur de thèse / Wissenschaftlicher Betreuer

Titlul tezei de doctorat ....

CONTRIBUTII LA TEORIA FUNCTIONALEI DENSITATE CU APLICATII IN TEORIA REACTIVITATII CHIMICE SI A ELECTRONEGATIVITATII

Title of thesis / Intitulé du sujet de thèse / Titel ....

CONTRIBUTIONS WITHIN DENSITY FUNCTIONAL THEORY WITH APPLICATIONS IN CHEMICAL REACTIVITY THEORY AND ELECTRONEGATIVITY

Data susținerii tezei de doctorat .... 15.03.2002 ....
Defended on / Date de la soutenance / Datum der Verteidigung

Rector,
Rector / Président de l'Université / Rektor

Secretar şef,
Registrar / Secrétaire général / Chefsekretär

Preface

The present English *Thesis* edition refines and enlarges the Romanian *Thesis* of the author, publicly defended in 15.03.2002 at West University of Timisoara, under the advisory of Prof. Dr. Eng. Adrian Chiriac. Since that date the author undertook an intensive Post-Doctorate research stage at University of Calabria under the advisory of Prof. Dr. Nino Russo. During this period the main results obtained in a theoretical fashion in author's *Thesis* were further extended and applied. For this reason, the present work likes also to emphasize, when opportunely, the fruitful prospective and aspects that emerged out from a certain author's original *Thesis* contribution, being no part of this work in conflict with other ventures currently under publishing considerations. However, by supplying the *Thesis* basic achievements with new ones will not affect the *Thesis* flavor but only enrich it.

Mihai V. Putz, Ph.D.

University of Calabria, Italy

Spring, 2003

Dedicated

to

Mama Kathy

and to

Papa Vio

My Steps, Yours Shines

Abstract

In the limits of the density functional theory there are introduced and deduced fundamental chemical descriptors as the chemical action concept, the chemical field, new electronegativity, rate reaction and chemical hardness formulations, the reduced total energy and the partial Hohenberg-Kohn functionals.

For electronic density computations the quantum statistic picture of the path integral Feynman-Kleinert formalism is employed to its markovian approximation, providing the framework in which the majority of the chemical reactions and the reactivity of the electronic systems can be treated together with the internal and environmental couplings.

Evaluation, representation and interpretation of the present analyzed chemical indices are performed for a prototype many-electronic system such that its electronic structure to display fundamental and excited anharmonic vibrations being in the thermal coupling with the medium.

The chemical descriptors introduced and computed shall contribute to the foundation of the chemical reactivity on the conceptual and analytical physical bases, being able to predict the chemical transformations and the characterization of the bonds formation.

Key Words:

Electronic Structure, Quantum Theory, Quantum Chemistry, Density Functionals, Reactivity Indices, Path Integrals, Feynman-Kleinert Formalism.

Acknowledgements

First of all, I would like to express my sincere gratitude to Professor Dr. Eng. Adrian Chiriac (Chemistry Department, West University of Timisoara) for receiving this project very enthusiastically and supervising afterwards its materialization with a lot of patience, support and trust.

I specially thank Prof. Dr. Nino Russo (Chemistry Department, University of Calabria) for giving me the unique opportunity to enlarge the main theoretical results from this thesis with fundamental applications.

To Prof. Dr. Onuc Cozar and Prof. Dr. Vasile Chis (Physics Faculty, Babes-Bolyai University, Cluj-Napoca) I express my warmly regards for their sustain, referee and constructive remarks on this work.

I thank Prof. Dr. Branko Dragovich (Institute of Physics-Belgrade) for his open spirit in reviewing of this thesis.

I deeply thank to Prof. Dr. Hagen Kleinert and Dr. Axel Pelster (Free University of Berlin) for their continuous support, fruitful discussions and recommends.

I am also grateful to DAAD (Germany) and MIUR-UNICAL (Italy) for supporting my research stages in Berlin and Cosenza-Rende, respectively.

Finally, I wish to thank the editor Shereen Siddiqui for her invaluable help in publishing this Thesis Book.

Mihai V. Putz, Ph.D.

University of Calabria, Italy

Spring, 2003

Praise for Thesis

Dr. Adrian Chiriac, Full Professor of Chemistry:

… Mr. Mihai V. Putz comes with a consistent physical background in the modern quantum chemistry. Such formation fully qualify him to get inside into the Density Functional Theory, founded by the pioneeristic works of a physicist and a mathematician, Walter Kohn and John People, respectively, together awarded for this theory with the Nobel Prize in Chemistry on year 1998. …Mr. Putz's thesis gives a valuable contribution for conceptual ascribing of the electronegativity as the central measure of an electronic systems' reactivity. Throughout the new density functional electronegativity proposal by Mr. Putz's work, crucial reactivity indices like the chemical action, the chemical hardness as well as the energetic functionals are rigorously related and giving space to further developments. … For the computational methodology Mr. Putz had correctly used the path integral formalism in its Feynman-Kleinert picture that is a meaningful and a referentially solution of the electronic density implementation consistent with the density functional and the V-representability framework. This way, the Mr. Putz's thesis provides an elegant, conceptual and analytical tool allowing integrate multidisciplinary studies for a large class of many-electronic systems together with their internal and environmental markovian couplings. …

West University of Timisoara, Romania

March, 2002

***

Dr. Nino Russo, Full Professor of Chemistry:

The thesis of Mr. Mihai V. Putz deals with the fundamental problem of physical and chemical interactions in fermionic open thermodynamical systems. The thesis is original and deserves attention from the scientific community. The study approaches a theoretical direction in order to describe the chemical reactivity through chemical descriptors and reactivity indices inferred based on the description of quantum statistical electronic systems in the density functional theory. In this context, there are introduced and phenomenological evaluated the density functionals of electronic systems involved in interaction, exchange and transformation processes within the thermodynamical couplings with the environment. ... The advantage of Mr. Putz approach is that within the path integral formulation he succeed to comprise all the main information for the electronic systems in the external applied potential. This way, the present study combines the path integral and density functional formulations in an unitary way consistent with the Hohenberg-Kohn theorems ... and could significantly enlarge the reactivity indices theory and its applications.

University of Calabria, Italy

March, 2002

***

Dr. Onuc Cozar, Full Professor of Applied Physics:

...Mr. Mihai V. Putz's work is situated as contents in the framework of molecular quantum mechanics and quantum chemistry. ... Starting from the general equations for changing in energy and chemical potential for an open electronic system the author inferred new density functionals within the density functional theory limits: the chemical action, the chemical field and its period, an original electronegativity density functional formulation. ... As a note, it can be said that, a more emphasis for the own results would be welcome. ... There is remarked the very careful attention paid to presentation that turns the whole work into a substantial thesis but not harmful overcrowded. ... The stile of discourse is clear and concise with a structure that leads with a coherent comprehensibility of the main literature concepts in the field as well as the author's rigorously integrated and applied contributions. ...

Babes-Bolyai University of Cluj - Napoca, Romania

December, 2001

***

Dr. Branko Dragovich, Research Professor of Physics:

...According to my positive experience with the path integral method in foundation of quantum theory (quantum mechanics, quantum field theory and string theory) I think that this method should be applied to any particular physical and chemical quantum system. Theoretical foundations of open molecular electronic systems just start from Feynman's path integral method in this Ph.D. thesis of Mihai V. Putz. I find such approach well founded, very attractive and promising. Mr. Putz's thesis presents significant contribution to the theory of open electronic systems, exhibits author's inventivety and contains sufficient number of original results. ...

Belgrade Institute of Physics, Serbia

December, 2001

List of Main Symbols

Symbol(s)	Meaning(s)
g	anharmonic coupling parameter
j	charge current
$W[j]$, $W(j)$	charge field: functional, function
C_A	chemical action
ω_C; ω, Ω	chemical field; frequency: proper, trial
$\eta(x), \eta_S, \eta_\chi$	chemical hardness: local, global from softness and global from electronegativity
$s(x,x')$, $s(x)$, S	chemical softness: kernel, local, global
A	classical action
ρ, $\rho(x)$	electronic densities
$\chi, \chi_M, \chi_P, \chi', \chi^{\sqrt{}}$	electronegativity: absolute, Mulliken, universal Parr, mean arithmetic and mean geometric
E, $E[\rho]$, ΔE	energy: total, functional, transferred
x_0	Feynman-centroid
a^2	Feynman-Kleinert smeared out width
$\overline{F}$, F	Feynman-Vernon functional, free energy
$f(x)$	Fukui function
$F_{HK}[\rho]$, $F_{HK}^{P}[\rho]$	Hohenberg-Kohn functional: universal, partial
β	inverse thermal energy
$L(x)$, $\dot{L}(x,x,t)$	local linear response function, the Langranjean
Z^*	nuclear charge
N, $N[\rho]$, ΔN	number of electrons: total, functional, transferred
$\hat{H}, \hat{W}(t)$	operator: Hamiltonian, density
γ, ξ	parameter: scaling, orbital zeta
Z, Z_1	partition function: standard, first order approximation
$Dx(\ )$	path integral measure
$V(x)$, $V_{eff}(x)$, $W_1(x)$	potential: external, effective, first order approximation
$\Gamma[\rho]$; $\Gamma[]$	rate reaction functional; gamma Euler function
t, τ	time: real, imaginary
x; p	spatial vector position; impulse coordinate
$m_0, \hbar, k_B$	standard constants: electronic mass, Planck, Boltzmann
a, b	symbolic integrals
T; T_C	thermodynamical temperature; chemical period
k	wave vector modulus, force constant

List of Figures and Tables

Contents

CONTRIBUTIONS WITHIN

DENSITY FUNCTIONAL THEORY

WITH APPLICATIONS IN

CHEMICAL REACTIVITY THEORY

AND ELECTRONEGATIVITY

Cards suffuse, display abound,

How can they be turned to swans?

Robert G. Parr – Density Functional Theory

CHAPTER 1.

MOTIVATION

Don't be afraid, I will go away like an echo…

Garcia Lorca - Song without flowering

During the time, the nature sciences evolved under different theoretical shapes in order to causatively describe the objective-experimental reality, the evolution and interaction of species. [1-10]

Physics and Chemistry were in turn, but also correlated, in state to discover in a proper way the innovating kind of evolution which characterizes the natural systems. Anyway, they set up the relative different ways of exploring, at various levels of interaction, the reality of manifestation of the same objects. The important thing is that the two approaches have always lend one to the other the major concepts (electron, atom, molecule, time dependence) in order to cooperate for the elucidation of the phenomenology of interaction and of the emergence of the natural (atomic, molecular, chemical, and biological) species in a self consistent organization. [5-11]

Since the appearance of the quantum mechanics, the (theoretical) methods of Chemistry and Physics used for describing the atoms and molecules made up an unit. [1, 2]

For instance, the description of the atomic and molecular systems through *the wave function*, beyond its abstract, mathematical and philosophical representation, that is still in the middle of enmities and interpretations, [2, 12] is like a bridge between the physical and chemical descriptions applied to the microscopic world. [13, 14]

However, besides the wave function (of atomic and molecular orbitals), other complementary (or alternative) models have been developed aiming to describe the multi-electronic systems. [15-23]

Such an example is the *density functional theory* (*DFT*), which has on the foreground the *electronic density* associated to a system in its fundamental quantum state. [18-20]

Certainly, also the quantum statistical models, based on *the partition function*, or the synergetic models, based on the *structure information*, have been developed, [7-9] but all of them can be reduced directly or indirectly to a correlation with the density or with the density of probability of the analyzed electronic system. Nevertheless, the big advantage of treating the multi-electronic systems through the associates density functionals consists in the *observable* character these functionals receive, once they are expressed as functions of electronic density. Moreover, the calculus of the electronic density can also be made on the base of the partition functions which, at their turn, can be calculated throughout an alternative formalism of solving the Schrödinger equation for the electronic systems. [15]

Such an alternative method is made up by the path integral(s)-*PI* formalism, presented both in its general quantum mechanical (*QM*) manner as well as in the quantum statistical (*QS*) realization within Feynman-Kleinert approach, [24] in the Chapter 2.

Unlike solving the Schrödinger equation, that assumes between two quantum events the quantum evolution gap, the calculus based on the path integrals takes into account, in principle-precisely, the amount of all possibilities of filling the quantum gap separating two arbitrary quantum events. Although these models are different, they coincide with a precision of under 1% for the electronic systems governed by the polynomial potentials (like the electronic molecular ones). [25]

The advantage of applying the path integrals formalism presents, at least, two correlated aspects. First of all, the partition function can be calculated for the (thermodynamically) open electronic systems, and moreover, this calculation (although presumes in most of the cases the effective potential approximation) doesn't imply a perturbation method, as is the case of the Schrödinger-Rayleigh standard approach. [14, 26-31]

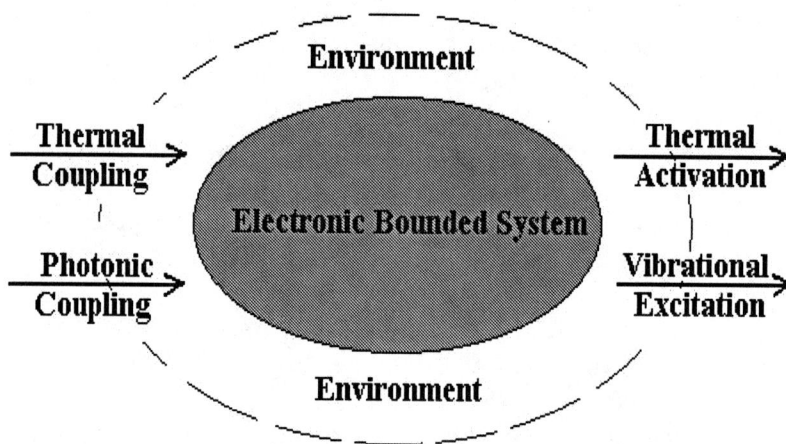

Figure 1.1. *The paradigm of the chemical-physical processes for an open electronic system.*

Being *DFT* a manifestly *structure* theory, due its bases on the stationary principles of the ground-state, its combination with *PI* quantum statistical scheme seems more adequate than with those of wave function. This because, the *QM* wave function does not provide directly the structure (electronic density) information, being in a multi-coordinate many-particle space dependence, while *QS* directly associates with the electronic density throughout the partition function, a global energetic measure of the structure.

In these conditions, to take into account the interactions, the changes and the couplings with the external environment it appears more accessible (for theoretical analysis) through the using of the path integrals method in solving the quantum statistic partition function – and then, the density associated to the open electronic systems. [32-38]

The idealization of the real systems is generally admitted, Figure 1.1. In fact, the evolution of natural systems (or species) cannot take place in the absence of the couplings, interactions, and exchanges between these systems and the external environment (seen as a collection of other open systems). Without emphasizing here on the physical and philosophical side of this statement - the Mach's principle, the principle of non-separability, and so on, [2, 4, 12] - one shall limit upon the importance of studying the open electronic systems, and of their exchanges and couplings with the external environment .

This work will analyze the potency of an termodyanamically (from here implicitly) open atomic and molecular electronic system to enter in a reaction channel. Beside the coupling with the nuclear system and/or with an external (photonic) electromagnetic field, the entire electronic ensemble is placed within a thermal bath allowing so far the thermal exchange, being the springing effect the excitation on the anharmonic vibrational states. [39-42]

The present work is structured in two main parts.

In Chapter 2 there are introduced and analytically deduced, within the limits of *DFT*, the chemical descriptors – as density functionals – with validity (also) for the open electronic systems: *the chemical action, the chemical field, the chemical field period, electronegativity (also with its Mulliken version and the atomic path integral sketched scale), new reaction rate, the reduced total energy and the partial Hohenberg-Kohn functionals.* All these descriptors are directly or indirectly related, by means of conceptual quantum chemistry, with the driving electronegativity as the systems' electronic potency to enter into a (virtual) reaction channel. [43-56]

In Chapter 3 are calculated, represented and analytically interpreted the chemical descriptors introduced in Chapter 2 for a prototype molecular electronic system, governed by a generalized anharmonic potential, employing the Markovian approximation of the ultra-short correlations with the environment. The calculus of the associated electronic densities emerges out from the method of the Feynman-Kleinert variational algorithm in *QS* path integrals. [57-63]

The chemical descriptors introduced in the present work show, through their density functional structure associable also to the open systems, a general character for the capacity of an electronic system to participate at a reaction. By their definition and the manner in which they were inferred, the considered chemical descriptors allow a wide range of density implementations, associated to the aimed electronic states, leading as well a coherent phenomenology of the reactivity.

This way, the present study proposes itself to contribute to fundament the chemical reactivity on physical bases.

CHAPTER 2.

THE PHYSICAL BASES OF THE CHEMICAL REACTIVITY

Prometheus: Release me, Zeus, I've already suffered enough...

Luciano di Samosata - Dialogs of Gods,

Prometheus and Zeus, 1

2.1 INTRODUCTION

The present chapter likes to formulate some chemical descriptors in a unitary way in the frame of the density functional theory (*DFT*) allowing so far a high level of generality which should include the case of open electronic systems too.

For instance, there have been formulated different manners of determination and measurement for electronegativity. [64-80] Although this descriptor is a fundamental one in characterization of charge exchange for forming and transforming of atoms and molecules, an analytical general expression of it, as density functional, hasn't yet been formulated. Based on the density functional theory, in this chapter, will be deduced an electronegativity density functional and its Mulliken version.

Equally, are defined and deduced other basic functionals, like is the case of Hohenberg-Kohn universal functional, presented here in its partial venture.

Considering the equations for the transformation of electronic ground states, within the density functional theory framework, new reactivity indices – as quantitative density functional descriptors of reactivity potency – will be introduced: the chemical action, the chemical field and its period, new rate reaction and chemical hardness functionals.

In the literature study, there is also presented the quantum mechanical and the quantum statistical respectively descriptions for the electronic systems' evolutions by means the path integrals perspective.

The quantum statistical path integral formulation is chosen as the main tool for the analytical calculus of the electronic densities, and will be fully employed in Chapter 3 for a prototype anharmonic molecular system, to compute the reactivity descriptors introduced in the present chapter.

All these considerations, formulations and deductions for defining and introducing of the reactivity density functionals – all from the direct or indirect electronegativity perspective – like to contribute for the characterization of an electronic system's reactivity through the adequate descriptors, displaying therefore a general predicting character, with the goal in providing an unitary view regarding the capacity of the electronic systems to enter in chemical reactions.

2.2 LITERATURE STUDY

2.2.1 The Electronegativity Concept

Historically, the electronegativity (χ) accounts for the tendency of the atoms to build up molecular systems. It was firstly proposed by J. J. Berzelius in 1811.

Linus Pauling in 1932, [64] by an ingenious mixture of quantum mechanical and thermodynamical arguments correlates electronegativity with the bond energy D of an arbitrarily species AB, A_2 and B_2 as:

$$\chi_A = \chi_B + 0.208 \left\{ D(AB) - \frac{1}{2}[D(A_2) + D(B_2)] \right\}^{1/2} [Energy]^{1/2}.$$

(2.1)

This way, it was introduced so far the referential electronegativity scale: once being fixed the χ of a chosen atom the other atomic values will depend of this choice according with relation 2.1.

Immediately after, Mulliken, during the years 1934-35, had proposed the electronegativity as the average of the binding energy of the outer electrons between the neutral atom (A) and its corresponding negative ion (A$^-$): [65]

$$\chi_M(A) = \frac{IP(A) + IP(A^-)}{2} \quad [Energy]$$

(2.2)

being IP the respective ionization potentials. After many years, it have been proved the deeply quantum nature of the Mulliken proposed χ formulation, [18] since the ionization potential was quantified by the integral of the highest occupied (molecular) orbital (*HOMO*) energy, ε_{HOMO}, respecting to the electronic population n of a given N-electronic system, $N = \sum_{orbitals} n_{orbitals}$:

$$-IP = E_N - E_{N-1} = \int_0^1 \varepsilon_{HOMO}(n)dn$$

(2.3)

being E_N and E_{N-1} the total energies of the N and $N-1$ electronic systems, respectively.

Afterwards, Gordy in 1946 finds a correlation of electronegativity with the effective nuclear charge Z_{eff} acting on the outermost electrons: [66]

$$\chi = \frac{Z_{eff}}{r_{cov}} \quad [Energy\,/\,electron]$$

(2.4)

establishing in such the (Coulombian) potential nature of electronegativity.

Iczkowski and Margrave in 1961, [67] are the first ones that have been made the correlation of electronegativity with the slope of the total energy curve $E(q)$:

$$\chi = -\left.\frac{\partial E}{\partial q}\right|_{q=N-Z*} \quad [Energy\,/\,electron]$$

(2.5)

with Z^* the total nuclear charge. This formulation also preserves for χ its previously established potential nature.

The definitive consecrated quantitative fashion of electronegativity arises in the framework of Density Functional Theory – DFT, by the work of Parr *et al.* in 1978, that fully identifies the electronegativity with the negative of the chemical potential: [72]

$$\chi = -\mu = -\left(\frac{\partial E}{\partial N}\right)_V \quad [Energy\,/\,electron]$$

(2.6)

being V the external (Coulombian, but not only) potential.

The last definition as simply was formulated as the efficient it is.

For instance, the previous association is even more natural as long the electronegativity is understood like a virtual measure in "seeing" electrons, and the chemical potential is regarded as the potency of "launching" electrons.

From the above two "contrary senses", but on the same "phenomenological direction" their (absolutely) identification unfolds naturally.

Remarkably, the ultimate Parr electronegativity definition, in the same time, correlates directly with energy, accounts for the number of electrons, and has the significance of the chemical potential of the system, preserving so its potential nature.

Moreover, when the parabolic dependence of the total energy E with the total number of electrons is assumed, see Figure 2.1, the finite difference approximation of relation 2.6 can be applied:

$$\chi = -\left(\frac{\partial E}{\partial N}\right)_V$$

$$\approx -\frac{E(N+h) - E(N-h)}{2h}$$

$$\overset{h=1}{=} \frac{-E(N+1) + E(N-1)}{2}$$

$$= \frac{[E(N-1) - E(N)] + [E(N) - E(N+1)]}{2}$$

$$= \frac{IP + EA}{2}$$

(2.7)

arriving back to the operational Mulliken χ definition, relation 2.2, being $EA = E(N) - E(N+1)$ the electron affinity of the N-electronic system. [81]

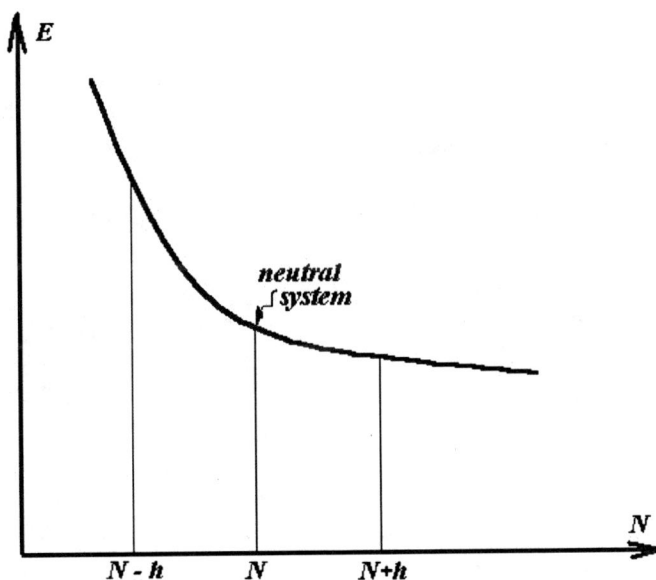

Figure 2.1. *Schematically representation of the parabolic dependence of the total energy of an N-electronic system, connecting its ionic (± h) states .*

However, worth to formulate an electronegativity expression, starting from Parr definition 2.6 and recovering the Mulliken 2.7 one, without excursion to the knowledge of the total energy dependence in the total number of electrons. Such a way is to be presented in the forthcoming sections of this chapter.

Nevertheless, being noted the importance to assign numbers to electronic systems starting from the atomic level, i.e. to quantify the electronic systems, worth to summarize the main features that any χ good proposed formulation have to satisfy: [82]

(i) χ must have a quantum mechanically viable definition;

(ii) χ must have the potential nature, i.e. to be an energy per electron, measured in electron-volts (eV).

At the atomic level, but not only, electronegativity can be adopted as the primary measure of the electronic stability of the systems.

Therefore, the electronegativity can provide a good qualitative and quantitative input for the elucidation of the binding processes, energy and charge exchanges between electronic systems.

The first and the most spread level of combined electronic systems regards the approached atoms to form molecules. This case, from the electronegativity perspective, is analyzed in the next section.

2.2.2 The Electronegativity of Atoms in Molecules

Beyond analyzing the atomic and the molecular orbitals, appears the necessity of a global characterization of the electronic distribution in a system (atomic, ionic, molecular or radicals) as bearer in a specific combination with other systems, defining the direction of reactivity's evolution, of charge transfer and of mutual change of properties.

Such a primary systemic characteristic is the electronegativity. Although many controversies upon this concept have raised, regarding its nature and namely what it really characterizes, ultimately there has been established a significant series of principles and characteristics that, qualitatively and quantitatively, correlate the atomic and molecular structures. More than the static definition that appeals the energy only, the electronegativity has to be seen more like a potential under which action the electrons can be exchanged with the environment. In this dynamic framework, the picture of the forming molecules by atoms measures the decline in total energy of the atoms that acquires electrons, exceeding the rise of respective atomic total energy. [83-89] The atomic properties in the molecule will be indicated by $\langle \, \rangle$ whereas the isolated atomic properties will not display such embracement.

The starting point is to consider the expansion of the atomic energy of an atom, around its N-electronic isolated ("0") status, up to the second order, when attempts to a molecular coordination throughout the charge transfer (ΔN):

$$E_{\langle\rangle}(N_{\langle\rangle}) \cong E_{\langle\rangle}(N) + \left(\frac{\partial E_{\langle\rangle}}{\partial N_{\langle\rangle}}\right)_0 (N_{\langle\rangle} - N) + \frac{1}{2}\left(\frac{\partial^2 E_{\langle\rangle}}{\partial N_{\langle\rangle}^2}\right)_0 (N_{\langle\rangle} - N)^2$$

$$\equiv E_{\langle\rangle}(N) - \chi\Delta N + \eta(\Delta N)^2$$

(2.8)

where the second order derivation of the total energy respecting with the number of electrons was identified as the first order electronegativity derivation respecting the number of electrons defining the *chemical hardness*: [81]

$$\eta \equiv \frac{1}{2}\frac{\partial^2 E}{\partial N^2}$$

$$= -\frac{1}{2}\frac{\partial \chi}{\partial N} \ .$$

(2.9)

Following above convention, the relation between electronegativity of an atom in coordination and its isolated electronegativity and chemical hardness looks like:

$$\chi_{\langle\rangle} = -\frac{\partial E_{\langle\rangle}}{\partial N_{\langle\rangle}} = \chi - 2\eta\Delta N \ .$$

(2.10)

However, to pass from the atomic electronegativity to the molecular one a working principle is required. Due to the identification of electronegativity with the chemical potential, at all organization levels of electronic systems, it can be used the thermodynamically feature of the chemical potential. When systems with different chemical potentials are combined, they exchange the particles (charges, electrons) between them until their chemical potentials will equalize. In terms of electronegativity, the corresponding principle is noted as the Sanderson equalization principle of electronegativities and sounds like follows: [89] *for the molecules in their fundamental state, the electronegativities of different electronic regions in molecule – are equal.*

Nevertheless, at this point, a shortly revelatory discussion have to take place. [90-92] Let's consider that the molecule is seen as the total system *<A>+<B>* of the two binding regions: the dynamic acceptor region *<A>* and the dynamic donor region *<B>*. Obviously, the conservation in the total energy and number of electrons is preserved, throughout any kind of molecular partition in dynamic acceptor and donor regions:

$$E = E_{\langle A \rangle} + E_{\langle B \rangle},$$

$$N = N_{\langle A \rangle} + N_{\langle B \rangle}.$$

(2.11)

The regional electronegativities will respectively be given as:

$$\chi_{\langle A \rangle} = -\left(\frac{\partial E_{\langle A \rangle}}{\partial N_{\langle A \rangle}} \right)_{N_{\langle B \rangle}}, \quad \chi_{\langle B \rangle} = -\left(\frac{\partial E_{\langle B \rangle}}{\partial N_{\langle B \rangle}} \right)_{N_{\langle A \rangle}}.$$

(2.12)

Now, the electronegativity of the total system (molecule) is representable, according with the above Sanderson equalization principle of regional electronegativities, consecutively as:

$$\chi = -\left(\frac{\partial E}{\partial N} \right) = -\left(\frac{\partial E}{\partial N_{\langle A \rangle}} \right)_{N_{\langle B \rangle}} = -\left(\frac{\partial E}{\partial N_{\langle B \rangle}} \right)_{N_{\langle A \rangle}}$$

$$= \chi_{\langle A \rangle} - \left(\frac{\partial E_{\langle B \rangle}}{\partial N_{\langle A \rangle}} \right)_{N_{\langle B \rangle}} = \chi_{\langle B \rangle} - \left(\frac{\partial E_{\langle A \rangle}}{\partial N_{\langle B \rangle}} \right)_{N_{\langle A \rangle}}.$$

(2.13)

The result 2.13, however, seems to prohibit the equality between global and regional electronegativities:

$$\chi \neq \chi_{\langle A \rangle} \neq \chi_{\langle B \rangle}.$$

(2.14)

Beyond of the perplexity given by inequalities 2.14, this paradox is simple solved observing that:

$$\Delta\chi_{\langle B\rangle}^{\langle A\rangle} = \left|\chi_{\langle B\rangle} - \chi_{\langle A\rangle}\right|$$

$$= \left|\chi_{\langle A\rangle} - \chi_{\langle B\rangle}\right| = \Delta\chi_{\langle A\rangle}^{\langle B\rangle}$$

$$= \left|\left(\frac{\partial E_{\langle A\rangle}}{\partial N_{\langle B\rangle}}\right)_{N_{\langle A\rangle}} - \left(\frac{\partial E_{\langle B\rangle}}{\partial N_{\langle A\rangle}}\right)_{N_{\langle B\rangle}}\right|$$

$$= \left|\nu_{\langle A\rangle} - \nu_{\langle B\rangle}\right|$$

$$\equiv \Delta\chi$$

(2.15)

where the transfer potentials have been introduced:

$$\nu_{\langle A\rangle} = \left(\frac{\partial E_{\langle A\rangle}}{\partial N_{\langle B\rangle}}\right)_{N_{\langle A\rangle}},$$

$$\nu_{\langle B\rangle} = \left(\frac{\partial E_{\langle B\rangle}}{\partial N_{\langle A\rangle}}\right)_{N_{\langle B\rangle}}.$$

(2.16)

There is clear from 2.15 that now the Sanderson's principle of regional electronegativities equalization takes place when the transfer potentials 2.16 cancel each other or, in other words, when the zero minimum difference between regional electronegativities, within the bond, is achieved:

$$\Delta\chi \to \min \to 0.$$

(2.17)

Even more, the condition 2.17 leads just with the variational principle:

$$d\chi = 0 \qquad\qquad (2.18)$$

that have to characterize the most stable equilibrium molecular ground state. This way, the Sanderson principle is consistently with variational principle in terms of (molecular) electronegativity. However, there is also another great consequence of the Sanderson principle, starting from the 2.18 form. If condition 2.18 is regarded inversely, so to speak when $\Delta\chi$ is non-zero, the binding process is promoted, i.e. the *Principle of Frontier Electron Theory* follows: [18, 93-95] *of two different sites with generally similar disposition for reacting with a given reagent, the reagent prefers the one which on the reagent's approach is associated with the maximum response of the system's electronegativity (n. a.). In short,* $\Delta\chi$ *(n. a.) big is good.* (Parr and Yang: 1984, 1989). An illustrative example on how the electronegativity differences promotes the binding and molecular formation regards the next discussion.

Let's considering the formation of a diatomic molecule *AB* with constant atomic nuclear charges at the equilibrium separating distance R_{AB}. For an infinitesimal transfer of electronic charges between the molecule's atoms, $N_{\langle A\rangle} = N_A + dN_{\langle A\rangle}$ and $N_{\langle B\rangle} = N_B - dN_{\langle B\rangle}$, the variation in the total energy $E = E_{\langle A\rangle} + E_{\langle B\rangle}$ can be written as:

$$dE = \left(\frac{\partial E}{\partial N_{\langle A\rangle}}\right)_{N_{\langle B\rangle}, R_{AB}} (N_{\langle A\rangle} - N_A) - \left(\frac{\partial E}{\partial N_{\langle B\rangle}}\right)_{N_{\langle A\rangle}, R_{AB}} (N_{\langle B\rangle} - N_B)$$

$$+ \left(\frac{\partial E}{\partial R_{AB}}\right)_{N_{\langle A\rangle}, N_{\langle B\rangle}} dR_{AB}.$$

(2.19)

Since, in the fundamental equilibrium state, $\partial E / \partial R_{AB} = 0$, $dE = 0$, and counting also of the previous 2.15-2.18 phenomenology, the last equation can be reduced as:

$$\left(\frac{\partial E_{\langle A\rangle}}{\partial N_{\langle A\rangle}}\right)_{N_{\langle B\rangle}, R_{AB}} = \left(\frac{\partial E_{\langle B\rangle}}{\partial N_{\langle B\rangle}}\right)_{N_{\langle A\rangle}, R_{AB}}$$

(2.20)

from which, the principle of equalization of the atomic (bond) electronegativities in the formed molecule is assured. Next, by employing this principle on the atoms *A* and *B* in *AB*, through the 2.10 type equation, yields:

$$\chi_{\langle A \rangle} = \chi_A - 2\eta_A \Delta N = \chi_{\langle B \rangle} = \chi_B + 2\eta_B \Delta N = \chi_{AB}$$

(2.21)

from which, immediately follows the number of the transferred charges – with the expression:

$$\Delta N = \frac{\chi_A - \chi_B}{2(\eta_A + \eta_B)}.$$

(2.22)

Counting also on the 2.8 type equation, the released energy due to the previous charge transfer will be therefore successively written as:

$$\Delta E = (E_{\langle A \rangle} - E_A) + (E_{\langle B \rangle} - E_B)$$

$$= [-\chi_A + \eta_A \Delta N + \chi_B + \eta_B \Delta N]\Delta N$$

$$= [-\chi_B - 2\eta_B \Delta N - \eta_A \Delta N + \chi_B + \eta_B \Delta N]\Delta N$$

$$= [-\eta_B \Delta N - \eta_A \Delta N]\Delta N$$

$$= -\frac{1}{2}[\chi_A - \chi_B]\Delta N$$

$$= -\frac{1}{4}\frac{(\chi_A - \chi_B)^2}{(\eta_A + \eta_B)}.$$

(2.23)

It is clear now, from both relations 2.22 and 2.23, that the electronegativity difference is crucial for binding promotion and bonding formation, as well in terms of exchanged number of electrons as for the released total energy.

The average value of the equalized electronegativity of atoms in molecule can be obtained submitting 2.22 in 2.21 to give:

$$\chi_{AB} \equiv \overline{\chi} = \frac{\eta_A \chi_B + \eta_B \chi_A}{\eta_A + \eta_B}.$$

(2.24)

However, the present (often called Parr-Pearson) approach does not provide a direct evaluation of molecular hardness from the atomic ones, even a formula can be laid down from relation 2.24 if is assumed the same proportionality between electronegativity and hardness at both atomic and molecular levels:

$$\chi_A = \theta\, \eta_A$$

$$\chi_B = \theta\, \eta_B$$

$$\chi_{AB} = \theta\, \eta_{AB}$$

$$(2.25)$$

being θ a proportional constant.

Inserting the relations 2.25 in 2.24, it turns out for the average molecular hardness the result:

$$\eta_{AB} \equiv \overline{\eta} = 2\,\frac{\eta_A \eta_B}{\eta_A + \eta_B}.$$

$$(2.26)$$

However, one just notes that the assumptions made in 2.25 claims that all pairs electronegativity-hardness, either for atoms or molecules, are correlated by the same factor θ. Therefore, this treatment seems that doesn't properly correlates the average electronegativity $\overline{\chi}$ with the average of the chemical hardness $\overline{\eta}$ without such "universal" proportionality. Further expansions of this Parr-Pearson model were made, [96] but the difficulties concerning the inherent correlation between $\overline{\chi}$ and $\overline{\eta}$ still remain.

An improvement to unify the electronegativity and chemical hardness of atoms in molecule using an alternatively analytical geometrical approach is currently in progress. [97]

However, relation 2.23 permits the immediate recovery of the "Hard and Soft Acids and Bases" (HSAB) principle, [79, 81, 98-100] an important conceptual principle to treat the molecular binding and reactive processes.

Writing the energy variation at transfer under the following forms: [101]

$$\Delta E = \Delta\Omega_A + \Delta\Omega_B,$$

$$\Delta\Omega_A = -\frac{\eta_A}{4(\eta_A + \eta_B)^2}(\Delta\chi)^2,$$

$$\Delta\Omega_B = -\frac{\eta_B}{4(\eta_A + \eta_B)^2}(\Delta\chi)^2,$$

$$\Delta\chi = \chi_A - \chi_B$$

(2.27)

the optimal energetic transfer will imply, for instance, the minimization of $\Delta\Omega_A$ respecting η_A, in conditions in which $\Delta\chi$ and η_B are maintained constant.

Therefore, the condition to achieve the optimum transfer results to be:

$$\eta_A = \eta_B$$

(2.28)

implying the fact that *the species with a high chemical hardness prefer the coordination with species that are high in their chemical hardness, and respectively the species with low softness (the inverse of the chemical hardness) will prefer reactions with species that are low in their softness*. This way, the HSAB – Pearson principle, was established. [79, 81]

The above relations were inferred for the general case, for which the form of the chemical potential (respectively the electronegativity with changed sign) is not specified.

However, Parr and Bartolotti in 1982, [87] have proposed a sort of "universal" form for the chemical potentials, and respectively, for the electronegativities. This one should satisfy the principle of the atomic χ – equalization principle in molecule, but under the form of the geometric average, recognizing the Sanderson's principle viability also in this form.

This way, for the atoms A and B in the molecule AB at equilibrium the next relations apply:

$$\chi_{AB} = \chi_{\langle A \rangle} = \chi_{\langle A \rangle}(N_A + \Delta N),$$

(2.29)

$$\chi_{AB} = \chi_{\langle B \rangle} = \chi_{\langle B \rangle}(N_B - \Delta N)$$

(2.30)

that can equivalently be written like:

$$\chi_{AB}^2 = \chi_{\langle A \rangle}\chi_{\langle B \rangle}$$

$$= \chi_{\langle A \rangle}(N_A + \Delta N)\chi_{\langle B \rangle}(N_B - \Delta N) \; .$$

(2.31)

From the energy conservation reasons, a similar relation have to take place also in terms of isolated atoms:

$$\chi_{AB}^2 = \chi_A \chi_B \; .$$

(2.32)

By equating the right hand sides of equations 2.31 and 2.32 results the geometrical general form for the dependence of the atomic electronegativity in molecule:

$$\chi_{\langle \, \rangle} = \chi \exp[-\gamma \, (N_{\langle \, \rangle} - Z^*)]$$

(2.33)

being γ an exponential scaling parameter, with an universal character. In these conditions, applying the equalization principle between the atomic electronegativities in molecules, it is found the charge transferred, in the binding process of the AB molecule, $\chi_{\langle A \rangle} = \chi_{\langle B \rangle}$, with the expression:

$$\Delta N = -\frac{1}{2\gamma} \ln\left(\frac{\chi_B}{\chi_A}\right) \; .$$

(2.34)

The corresponding released energy, the analogue of the 2.23 relation, will be calculated now as:

$$- \Delta E = \int_{N_A = Z_A^*}^{N_A + \Delta N} \chi_{\langle A \rangle} dN_{\langle \, \rangle} + \int_{N_B = Z_B^*}^{N_B - \Delta N} \chi_{\langle B \rangle} dN_{\langle \, \rangle}$$

$$= \chi_A \frac{1}{\gamma} [1 - \exp(-\gamma \, \Delta N)] + \chi_B \frac{1}{\gamma} [1 - \exp(+\gamma \, \Delta N)]$$

$$\cong -(\chi_B - \chi_A)\Delta N - \frac{1}{2}(\chi_A + \chi_B)\gamma \, (\Delta N)^2 \; .$$

(2.35)

Note that in the relation 2.35, with 2.34, the limitation at the first order corresponds to the released energy given by equation 2.23.

From relations 2.9 and 2.33 automatically results also the expression for the γ parameter, namely:

$$2\eta_{()} = -\frac{\partial \chi_{()}}{\partial N_{()}}$$

$$= \gamma \, \chi \, \exp[-\gamma \, (N_{()} - Z^*)]$$

$$= \gamma \, \chi_{()};$$

$$\Rightarrow \gamma = \frac{2\eta_{()}}{\chi_{()}}$$

(2.36)

which gives the analytical way that permits the numerical evaluation of this constant for a given form of atomic energy.

Parr and Bartolotti, [87] have performed a general determination of this γ constant, by interpreting each process of coordination in terms of a series of atomic states through the fundamental ionic transformations:

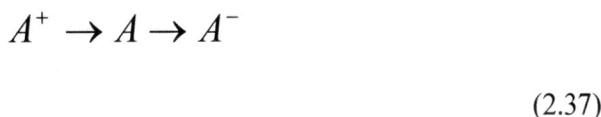

$$A^+ \rightarrow A \rightarrow A^-$$

(2.37)

being so admitted, as the basic coordination process, the addition and the promotion of an atomic electron in the molecular bond. As far as in the fundamental process 2.37 the total involved electrons varies from 0-to-2 it can be further considered the "frozen core" approximation, in which the atomic core is not participant. For such a situation a proper density description is provided by Slater atomic density orbitals, under the analytical form:

$$\rho(r,\xi) = N\frac{\xi^3}{\pi}\exp[-2\xi\, r]$$

(2.38)

being r and ξ the radial position vector of electrons in atoms and the orbital zeta parameter, respectively.

The density 2.38 obviously satisfies the *N*-normalization condition:

$$\int_0^\infty 4\pi\, r^2 \rho(r,\xi)dr = N$$

(2.39)

where the integration rule was used:

$$\int_0^\infty x^n \exp(-\alpha x)dx = \frac{n!}{\alpha^{n+1}}\,.$$

(2.40)

The orbital zeta parameter ξ in the expression 2.38 will be fixed by the minimization of the atomic promotion energy in the molecular bond:

$$\frac{\partial E[\xi]}{\partial \xi} = 0\,,$$

(2.41)

$$E[\xi] = T[\xi] + V_{ee}[\xi] + V_{ne}[\xi]\,.$$

(2.42)

The energetic components in 2.42 are the kinetic energy T, the electron-electron repulsion energy V_{ee} and the nucleus-electronic interaction energy V_{ne}, respectively. Their working formulations are:

$$T[\xi] = \int_0^\infty 4\pi\, r^2 \left\{ -\frac{1}{2}\left[\frac{1}{r^2}\frac{\partial}{\partial r}\left(r^2 \frac{\partial}{\partial r} \right) \right] + \frac{\xi}{2r} \right\} \rho(r,\xi)dr\,,$$

$$V_{ne}[\xi] = -\int_0^\infty 4\pi\, r^2 \frac{\rho(r,\xi)}{r}dr\,,$$

$$V_{ee}[\xi] = \frac{N-1}{2N}\iint \frac{\rho(1)\rho(2)}{r_{12}}dv(1)dv(2)$$

(2.43)

being counted in kinetic energy also the centrifugal orbital term.

The electron-electron repulsion term was written considering also the Fermi-Amaldi $(N-1)/N$ factor, [18] that assures for the correct self-interaction behavior: when only one electron is considered the self-interaction energy have to be zero: $V_{ee}(N \to 1) \to 0$.

The integrals of the first two formulas in 2.43 are already in radial coordinates while the last one, the electron repulsion integral, was let for the moment as the integration over the occupied volumes of the electrons 1 and 2. When the density 2.38 is substituted into the kinetic and nucleus-electron energetic terms of 2.43 one immediately gets:

$$T[\xi] = N \frac{\xi^2}{2},$$

$$V_{ne}[\xi] = -N\xi .$$

$$(2.44)$$

However, to evaluate the electron-electron repulsion energy, by using the density 2.38, much care have to be paid.

One has to use the electrostatic Gauss theorem which prescribes that the classical electrostatic potential outside an uniform spherical shell of charge is just what it would be if that charge were localized at its center and that the potential everywhere inside such a shell has the value at the surface, [13] see Figure 2.2.

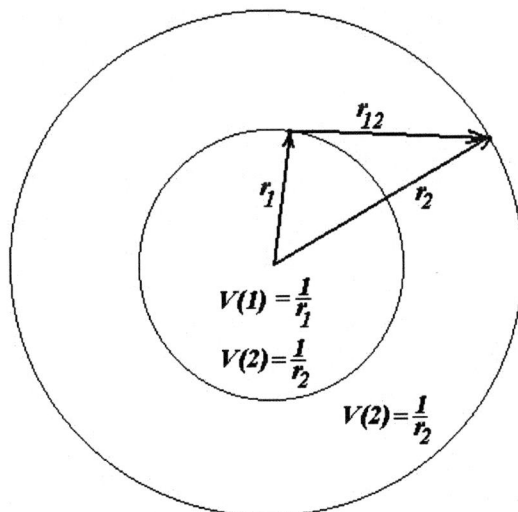

Figure 2.2. *Representation of the space regions of the 1 and 2 electrons, their potential influences and reciprocal interaction.*

Therefore, the electronic repulsion energy successively becomes:

$$V_{ee}[\xi] = \frac{N-1}{2N} \iint \frac{\rho(1)\rho(2)}{r_{12}} dv(1)dv(2)$$

$$= \frac{N-1}{2N} \int_0^\infty 4\pi \, r_1^2 N \frac{\xi^3}{\pi} \exp(-2\xi \, r_1)dr_1 \left[\int \frac{\rho(2)}{r_{12}} dv(2) \right]$$

$$= \frac{N-1}{2N} \int_0^\infty 4 r_1^2 N\xi^3 \exp(-2\xi \, r_1)dr_1 \left\{ 4\pi \, N \frac{\xi^3}{\pi} \left[\int_0^{r_1} \frac{r_2^2 \exp(-2\xi \, r_2)}{r_1} dr_2 \right. \right.$$
$$\left. \left. + \int_{r_1}^\infty \frac{r_2^2 \exp(-2\xi \, r_2)}{r_2} dr_2 \right] \right\}$$

$$= \frac{N-1}{2N} N^2 16\xi^6 \int_0^\infty r_1^2 \exp(-2\xi \, r_1)dr_1 \left\{ \left[\int_{r_2 \to r_1}^{r_2 \to \infty} \left(\frac{1}{r_1} \equiv \frac{1}{r_2} \right) r_2^2 \exp(-2\xi \, r_2)dr_2 \right. \right.$$
$$\left. \left. + \int_{r_1}^\infty r_2 \exp(-2\xi \, r_2)dr_2 \right] \right\}$$

$$= \frac{N-1}{2N} N^2 32\xi^6 \int_0^\infty r_1^2 \exp(-2\xi \, r_1)dr_1 \left[\int_{r_1}^\infty r_2 \exp(-2\xi \, r_2)dr_2 \right]$$

$$= \frac{N-1}{2N} N^2 \frac{32\xi^6}{(2\xi)^5} \int_0^\infty (2\xi \, r_1)^2 \exp(-2\xi \, r_1)d(2\xi \, r_1) \left[\int_{2\xi \, r_1}^\infty 2\xi \, r_2 \exp(-2\xi \, r_2)d(2\xi \, r_2) \right]$$

$$\equiv \frac{N-1}{2N} N^2\xi \int_0^\infty s^2 \exp(-s) \left[\int_s^\infty t \exp(-t)dt \right] ds$$

$$= \frac{N-1}{2N} N^2\xi \int_0^\infty s^2 \exp(-s) \left[(1+s)\exp(-s) \right] ds$$

$$= \frac{N-1}{2N} N^2 \xi \left(\frac{2!}{2^3} + \frac{3!}{2^4} \right);$$

$$\Rightarrow V_{ee}[\xi] = (N^2 - N) \frac{5}{16} \xi .$$

(2.45)

Summing up the energetic terms from 2.44 and 2.45, applying the optimum condition 2.41, and solving the obtained equation for the ξ, the solution looks like:

$$\xi = \frac{21 - 5N}{16} .$$

(2.46)

With the expression 2.46, back in the energetic terms 2.44 and 2.45, the partial derivatives of the total energy 2.42 respecting to the total number of electrons can be taken, aiming to determine the corresponding electronegativity 2.6 and the chemical hardness 2.9. For the imposed condition $N=1$, according to the type of process 2.37, there are obtained the respectively values:

$$\chi_{(N=1)} = -\frac{\partial E}{\partial N}\bigg|_{N=1} = \frac{3}{16}(\text{a.u.}) = 5.1(\text{eV}),$$

(2.47)

$$2\eta_{(N=1)} = \frac{\partial^2 E}{\partial N^2}\bigg|_{N=1} = \frac{135}{(16)^2}(\text{a.u.}) .$$

(2.48)

Form these expressions result two types of constants which can be considered representative, or "universal", for their value.

Equation 2.47 states as a reasonable kind of universal atomic electronegativity around the value of 5.1 eV, named hereafter universal atomic Parr electronegativity.

Moreover, from 2.36, with the help of 2.47 and 2.48, results the value of the searched γ parameter:

$$\gamma = \frac{2\eta}{\chi} = 2.8 \qquad (2.49)$$

with which help, the exchange number of electrons 2.34 and the respectively released energy 2.35, for different species of atoms, can be evaluated within the geometrical mean molecular binding approximation. However, an essential aspect can be remarked related to the characterization of the chemical bond throughout electronegativity, namely that it has always a companion: the chemical hardness. The deeply understanding of the relation between these two descriptors of the chemical bond can be formulated only within the framework of a general theory, as the density functional theory is.

This aspect, among a wide theoretical frame, in which the chemical bond is seen like a part of the binding process, will be axiomatically presented in the next two sections.

2.2.3 Theorems of Density Functional Theory and Consequences

Quantum physicists and chemists strive to discover theoretical formulations for understanding and compute the structure of matter and its modifications. A modern tool for this endeavor is the density functional theory (*DFT*), [16-23] that presents the immense advantage of treating the many-body systems in terms of the single effective particle density, that it is, in principle, exact. "A great strength of the density theory functional language is its appropriateness for defining and elucidating important universal concepts of molecular structure and molecular reactivity. In traditional quantum chemistry this has, of course, been a major goal, but it is tortuous to try to conceptualize how many body wave functions are related to structure and behavior. In *DFT* not only is the electron density itself very easy to visualize, but there is a big advantage that the electron number N has the central place in theory. After all, much of chemistry is about the transfer of electrons from one place to another." (Kohn, Becke, and Parr, 1996, [20]). Density Functional Theory is based upon two fundamental theorems called Hohenberg-Kohn (*HK*) theorems. [16]

(i)   The ground state electronic density $\rho(x)$ determines everything about a chemical system and integrates to the total number of electrons, N, in the system:

$$\int \rho(x)dx = N$$

(2.50)

being x the spatial position vector variable. More, the total energy functional of the system evolving under the external potential $V(x)$ separates as:

$$E[\rho] = F_{HK}[\rho] + C_A[\rho],$$

(2.51)

$$F_{HK}[\rho] = T[\rho] + V_{ee}[\rho].$$

(2.52)

The expression 2.52 states the *HK* functional as the sum of the electronic kinetic functional $T[\rho]$ and the electron repulsion functional $V_{ee}[\rho]$, whereas

$$C_A[\rho] = \int \rho(x)V(x)dx$$

(2.53)

means the *chemical action* term in 2.51 which will be also further analyzed.

The functional F_{HK} is generally not known but it has the quality that is universal: its functional form, as a functional of ρ, is independent of the external potential *V*.

Consequently, if F_{HK} is known for one particular external potential *V* and interaction potential V_{ee} it can be applied to another system with different external potential *V'* with the same interaction potential V_{ee}.

However, being $F_{HK}[\rho]$ an universal functional, the chemical action should make the difference between various chemical systems "from a small molecules to the giant protein" (Par and Yang, 1989, [18]).

More, the first *HK* theorem legitimizes the use of the density ρ as the basic variable instead of the number of electrons *N* or of the external potential *V*. Note that *V* is not restricted to the Coulomb potentials.

In other words, the external potential is determined, within a trivial additive constant, by the electronic density.

To sustain this last assertion, the second *HK* theorem is necessary.

(ii)   For any trial electron density $\bar{\rho}$ holds the variational principle in *DFT*:

$$E[\bar{\rho}] \geq E[\rho] \Leftrightarrow \delta E[\rho] = 0$$

(2.54)

around the ground state density ρ.

The consequences of these theorems are immense.

The first fundamental consequence of variational principle 2.54 consists in its equivalent stationary equation:

$$\delta\{E[\rho] - \mu N[\rho]\} = 0$$

(2.55)

where μ is the Lagrange multiplier associated to the fixed total number of electrons of the system.

Given 2.55, the μ multiplier, defined as the functional derivative:

$$\mu = \left(\frac{\delta E[\rho]}{\delta \rho} \right)_{\rho = \rho(V)}$$

(2.56)

for the correct density of the fundamental state, represents the chemical potential associated to the system that is further identified with the minus electronegativity of the system:

$$\chi = -\mu = -\left(\frac{\delta E[\rho]}{\delta \rho} \right)_{\rho = \rho(V)}.$$

(2.57)

To recover definition 2.6 from the density functional 2.57 the use of the first *HK* theorem is necessary, in the form 2.50, that provides the equivalences:

$$\chi = \chi \left(\frac{\partial N}{\partial N} \right)_V$$

$$= \int \chi \left(\frac{\partial \rho}{\partial N} \right)_V dx$$

$$= -\int \left(\frac{\delta E}{\delta \rho} \right)_V \left(\frac{\partial \rho}{\partial N} \right)_V dx$$

$$= -\left(\frac{\partial E}{\partial N} \right)_V$$

(2.58)

on the base of functional derivation rules. [18]

Nevertheless, the transformations 2.58 leads with another important consequence for reactivity in *DFT*, namely the fact that the equivalence between the functional electronegativity 2.57 and the standard electronegativity definitions 2.6 and 2.58 is assured throughout the function:

$$f[\rho_V,x]=\left(\frac{\partial\rho(x)}{\partial N}\right)_V$$

(2.59)

called Fukui function. The role of this function is of great importance in establishing the energetic site reactivity and will be further discussed in the subsequent sections.

As before stated, even implicitly enounced in first *HK* theorem, the one-to-one correspondence between ρ and V associated to the ground electronic state requires the use of the second *HK* theorem, so that it seems most appropriate for this demonstration to be placed as a consequence of both theorems. The demonstration uses the method of reduction *ad absurdum*.

Let's consider that the ground state density ρ is associated with two external potentials (V_1, V_2), and therefore with two Hamiltonians (H_1, H_2), two total energies (E_1, E_2) and two wave functions (ψ_1, ψ_2).

When ψ_1 is assumed as the right ground state wave function, the application of the variational principle 2.54 rewrites as:

$$E_1[\rho]=\langle\Psi_1|H_1|\Psi_1\rangle<\langle\Psi_2|H_1|\Psi_2\rangle$$

$$=\langle\Psi_2|H_2|\Psi_2\rangle+\langle\Psi_2|(H_1-H_2)|\Psi_2\rangle$$

(2.60)

leading, on the base of 2.51 and of the F_{HK} universality, with:

$$E_1[\rho]<E_2[\rho]+\int\rho(x)[V_1(x)-V_2(x)]dx.$$

(2.61)

On the contrary, when ψ_2 is assumed as the right ground state wave function the resulting inequality from variational principle 2.54, in this case, looks like:

$$E_2[\rho]<E_1[\rho]+\int\rho(x)[V_2(x)-V_1(x)]dx.$$

(2.62)

Summing up the relations 2.61 and 2.62 results the obvious contradiction:

$$E_1[\rho]+E_2[\rho]<E_1[\rho]+E_2[\rho].$$ (2.63)

Avoiding the false conclusion 2.63 means that to the ground state density ρ it corresponds an unique external (applied) potential V.

Now, the *HK* theorems propose the picture in which not the external potential but the electronic density determines all the properties of the system. This way, given a density, one has to be sure that it uniquely correlates with a potential fixing right the system's Hamiltonian which, generates the wave function giving the same proposed density. This is the condition for a density to be *V*-representable, a very difficult yet not solved task *per se*.

Fortunately, it turns out that the density functional theory can be formulated in a way that only requires that the density, both in functionals and in variational principle, satisfies a weaker condition, namely the *N*-representability one. The density ρ is said to be *N*-representable, when, among of the *HK* (*N*-integrability, or the proper normalization) condition 2.50, it fulfills also the non-negativity condition:

$$\rho(x) \geq 0 \ , \ \forall x \in \Re$$

(2.64)

and is a continuous non-divergent function on the real domain:

$$\int_{\Re} \left| \nabla \rho(x)^{1/2} \right|^2 dx < \infty .$$

(2.65)

Obvious, any reasonable density satisfy the *N*-representability conditions 2.50, 2.64, 2.65.

But how it can be by-passed the *V*-representability problem?

The response appeals the Levy's constrained search *DFT* formalism. [18]

The Levy's recipe prescribes that the ground state (*GS*) energy minimization procedure (within the second *HK* theorem) involves, in fact, two steps: one over all wave functions (ψ) that give the same density (the inner minimization step) followed by the minimization throughout all density classes (the outer minimization step):

$$E_{GS} = \min_{\rho} \left[\min_{\Psi \to \rho} \langle \Psi | (T + V_{ee} + V) | \Psi \rangle \right]$$

$$= \min_{\rho} \left[\min_{\Psi \to \rho} \langle \Psi | (T + V_{ee}) | \Psi \rangle + \int \rho(x) V(x) dx \right]$$

$$= \min_{\rho}\left(F_{HK}[\rho] + C_A[\rho]\right)$$

$$= \min_{\rho}\left(E[\rho]\right).$$

(2.66)

This way, the second *HK* theorem is detailed such that to assure the constrained search of the wave function able in providing the same density as that one corresponding to the external potential. This important consequence, and often seen as the Levy theorem – a corollary to the Hohenberg-Kohn ones, leaves a lot of freedom to practical density implementation in terms of *N*-representability instead of *V*-representability one. However, a careful account for constrained-searching condition, or equivalences, have to be considered. It will be shown in few sections that Path Integral Feynman-Kleinert formalism can provide a suitable equivalence in terms of partition function for the role wave function plays in the original Levy's procedure, as above described.

Another outstanding *DFT* consequence regards the chemical reactivity principles emerging out from the *HK* theorems. The chemical reactions involve the charge transfer. Therefore is compulsory to derive principles governing charge transfer in *DFT*. If one considers the fundamental relation 2.50 re-written as the density variation subject to constrain:

$$\int \Delta\rho(x)dx = \Delta N$$

(2.67)

then, the modification in the ground state energy, by means of the Taylor series expansion (assuming $\Delta\rho$ small enough to provide truncation to the second order), writes as:

$$E[\rho + \Delta\rho] \cong E[\rho] + \int\left(\frac{\delta E[\rho]}{\delta \rho(x)}\right)_V \Delta\rho(x)dx$$

$$+ \frac{1}{2}\iint\left(\frac{\delta^2 E[\rho]}{\delta \rho(x)\delta \rho(x')}\right)_V \Delta\rho(x)\Delta\rho(x')dxdx'.$$

(2.68)

Next, is required that the deviation of energy $E[\rho + \Delta\rho]$ to be minimum. What does mean this in terms of reactivity? It means that the active sites of a reactant molecule are usually placed where the addition or loss of electrons is energetically favorable. Back into mathematical language, the most favorable sites to add or lose electrons further mean that the energy 2.68 have to be minimized by a function that just accounts for the ground state density variations when electrons are exchanged; this one identifies with the previous introduced Fukui function 2.59. Thus, the inserting presence of the Fukui function 2.59 into 2.68 will produce the minimization of the changed energy, by accounting for the preferred reactive sites, leading the analytical successive equivalences:

$$\Delta E \cong \int \left(\frac{\delta E[\rho]}{\delta \rho(x)} \right)_V \left(\frac{\partial \rho(x)}{\partial N} \right)_V \Delta \rho(x) dx$$

$$+ \frac{1}{2} \iint \left(\frac{\delta^2 E[\rho]}{\delta \rho(x)\delta \rho(x')} \right)_V \left(\frac{\partial \rho(x)}{\partial N} \right)_V \left(\frac{\partial \rho(x')}{\partial N} \right)_V \Delta \rho(x) \Delta \rho(x') dx dx'$$

$$= \left(\frac{\partial E}{\partial N} \right)_V \Delta N + \frac{1}{2} \left(\frac{\partial^2 E}{\partial N^2} \right)_V (\Delta N)^2$$

$$= -\chi (\Delta N) + \eta (\Delta N)^2 .$$

(2.69)

In deriving 2.69, the functional derivative rules like in deriving 2.58, together with the standard definitions of electronegativity 2.6, 2.58 and of the chemical hardness 2.9, as well as the constraint 2.67, were considered. However, the result 2.69 enlightens two major achievements. The first one states that the electronegativity and chemical hardness are the minimizing global values corresponding to the local Fukui minimizing function for the change in total energy functional, when the system is favorable to exchange electrons around its ground state. The second result shows that for enough small variation in the ground state density the shape of the change in energy displays a parabolic dependence on the exchanged number of electrons. So, the parabolic assumption for the total energy in number of number of electrons, even not demonstrated to be most generally valid, finds however a natural justification, and fits with a wide range of the electronic exchange processes.

It remains to explore the reverse problem, that is: how are optimized the values χ and η during a chemical reaction or through the energy minimization processes? To give a reasonably answer, firstly consider the parabolic shape of the total energy respecting with the total number of electrons, as in Figure 2.1, connecting the three states of interests: $|N-h\rangle$, $|N\rangle$, and $|N+h\rangle$ ones.

Let's next assume that any energetic shape, containing the associated energies for the above considerate electronic states, has a parabolic form, as certifying by relation 2.69, for instance.

Then, one is interested in minimizing the energy throughout all parabolic classes that link those states. Is immediately that such a minimizing procedure can be undertook in two distinct ways: to simultaneously minimize the energetic values of the states $|N-h\rangle$ and $|N+h\rangle$, or to only act on the energy in the "point" $|N\rangle$ of the energetic shape, as in cases *(I)* and *(II)* in Figure 2.3, respectively. [102]

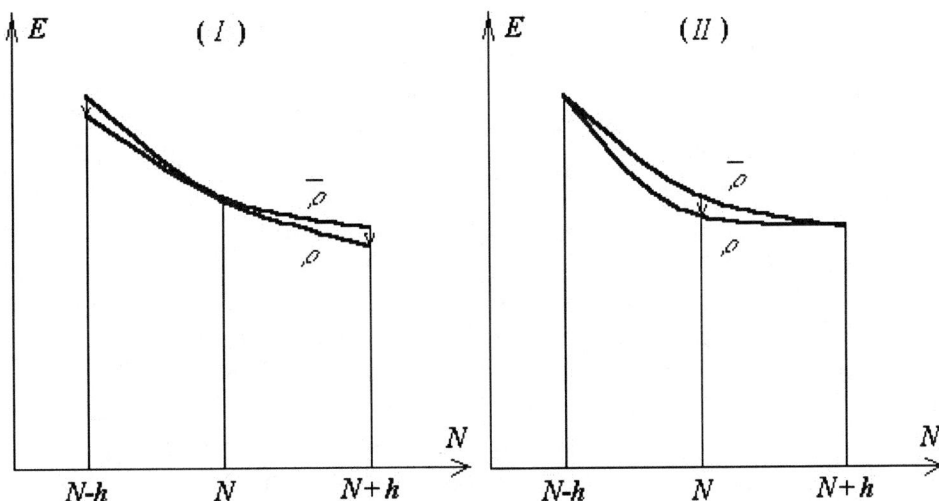

Figure 2.3. *The two cases of the total energy minimization by distinct acting on the parabolic energetic shape connecting the electronic states* $|N-h\rangle$, $|N\rangle$, *and* $|N+h\rangle$.

In the representations *(I)* and *(II)* of Figure 2.3 the negative slope and the convexity of the energetic shape for the state $|N\rangle$ will give information about the behavior of χ and η during the energy minimization, respectively. Firstly, in both analyzed cases the electronegativity approaches its minimum on the right ground density ρ, around the electronic state $|N\rangle$:

$$(I),(II): \overline{\chi}\,[\overline{\rho}] \geq \chi\,[\rho]$$

(2.70)

whereas the chemical hardness records different optimum values depending with the type of energetic minimization procedure:

$$(I) : \bar{\eta} \, [\bar{\rho}] \geq \eta \, [\rho],$$

$$(II) : \bar{\eta} \, [\bar{\rho}] \leq \eta \, [\rho] .$$

$$(2.71)$$

Now arises the dilemma: what kind of optimization behavior is the correct one for the chemical hardness, a minimum value, as in case (*I*) or a maximum one like in case (*II*), from Figure 2.3, for the right ground ρ state, respectively? There are two arguments to chose that the right ρ ground state have to have the maximum hardness value, at the end of the energetic minimization (binding) process. The first argument calls the relation between electronegativity and hardness, trough the relation of type 2.9, that prescribes the negative sign between the electronegativity and chemical hardness tendencies when the right ground state is approached. Now, being associated the right ground state, in both above (*I*) and (*II*) cases, with the minimum value in electronegativity, see 2.70, the natural choice for the chemical hardness for ending the minimization process will be therefore its maximum value. The second argument regards the parabolic energetic shape, that from beginning it was assumed to be maintained as the invariant characteristic during the minimization process. And is clear that in the case (*I*) of Figure 2.3 the minimization of energy tends to deform the parabolic energetic shape into a linear one. Thus, the right way of energy minimization trough the parabolic classes corresponds with the case (*II*) which, in turn, prescribes the maximum hardness optimization for the achievement of right ground state, see 2.71. Ending the last issue and this section, worth noting that Pearson had pointed out that *there seems to be a rule of nature that molecules* (or the many-electronic systems in general, n. a.) *arrange themselves* (in their ground state, n. a.) *to be as hard as possible*. [79] This way, was established the Maximum Hardness Principle (*MHP*) for the equilibrium of the many-electronic systems in their ground states. [103-107]

2.2.4 Reactivity Indices in Density Functional Theory

"The ultimate goal of any general theory of chemistry must be to give information about the relative stabilities of molecules, and their tendencies to undergo chemical change under specific conditions" (Pearson, 1997, [107]).

Indeed, once presented the fundament of *DFT*, as a new general theory of chemistry, it has to be further applied for description of molecular stability and reactivity. [108-127]

Before to further proceed let's give space, again, to the thoughts of one pioneer not only of quantum chemistry itself, but also of the Density Functional Reactivity Theory, Robert G. Parr: "we...demonstrate (that *DFT* has an, *n. a.*) unique power by considering the equations governing the change from one ground to another. ... A soft atom or molecule generally has a high propensity to form covalent bonds, and reactivity indices of the frontier type generally identify sites where chemical bonds are likely to form. One therefore expects the hardness-softness ideas to be intimately related to the reactivity index ideas. We here explore this relationship, and in the process discover natural definitions for local softness, local hardness, and softness and hardness kernels. ... A splendid relevance for chemistry will be revealed." (Parr and Yang, 1989, [18]).

The most general differential equation for the change in the functional energy $E=E(N,V)$ is taken as:

$$dE = -\chi\, dN + \int \rho(x) dV(x) dx$$

(2.72)

with the electronegativity χ being given by the already introduced relation 2.6 and where, it was recognized also the relation:

$$\rho(x) = \left(\frac{\delta E}{\delta V(x)} \right)_N$$

(2.73)

through the formal functional differentiation of 2.51-2.53 and accounting for the universal nature (not the V - dependency) of the Hohenberg-Kohn functional F_{HK}. Equation 2.72 states as the *first equation of the chemical physical processes* in *DFT*, because it links the total energy variation of an electronic state (atomic or molecular) to the charge exchange (dN) at which is subjected, and to the potential variation (dV) which acts on the initial state. Applying the Maxwell relation on the total differential equation 2.72 results an identity which involve the Fukui function $f(x)$, see 2.59, and thus redefining it as:

$$-\left(\frac{\delta \chi}{\delta V(x)} \right)_N = \left(\frac{\partial \rho(x)}{\partial N} \right)_V = f(x) \ .$$

(2.74)

The new definition 2.74 of the Fukui index better reveals its reactivity nature as a descriptor which measures how sensitive a system's electronegativity is to an external perturbation at a particular point x.

Analogous, it can be expressed the total differential of electronegativity as a functional of the number of electrons in the system and of the applied external potential:

$$ d\chi = \left(\frac{\partial \chi}{\partial N} \right)_V dN + \int \left(\frac{\delta \chi}{\delta V(x)} \right)_N \delta V(x) dx \ . $$

(2.75)

The first term on the r.h.s. of the equation 2.75 it can be recognized as the chemical hardness associated to the electronic system, according to the definition 2.9, whereas for the second r.h.s. term the identities 2.74 can be applied. The resulting electronegativity differential equation in terms of chemical hardness and Fukui function reads as:

$$ d\chi = -2\eta \, dN - \int f(x) dV(x) dx \ . $$

(2.76)

This 2.76 equation states as *the second equation of the chemical physical processes* in *DFT*, equivalent to the first one, equation 2.72, but rewritten at another level. Explicitly, equation 2.76 correlates the exchange of the electronegativity of an electronic state (atomic or molecular) with the charge exchange and the potential modification throughout the chemical hardness and the Fukui frontier function, respectively, leading with the electronegativity driving role for the variation in the frontier orbital charge. [20] However, worth to point out other important features of equation 2.76. Firstly, it contains explicitly the variables N and V, the central ones in *DFT*. The number of electrons N encompasses the many-body approach, whereas V fixes the equilibrium ground state of those N electrons to whom applies. Thus, only reactivity indices are appearing in 2.76, without explicitly referee to the knowledge of the total energy and of the Hohenberg-Kohn functionals, $E[\rho]$ and $F_{HK}[\rho]$, respectively.

Therefore, finding a suitable $[N, V]$ formulation of the indices χ, η, and $f(x)$ will finally lead with a proper density functional reactivity representation.

An equivalency of the *DFT* variation principle 2.54, or 2.55, can be formulated throughout the relation 2.56 combined with the axiomatic *DFT* prescriptions, equations 2.51-2.53, so providing the Euler-Lagrange equation of the electronic system in terms of electronegativity:

$$ -\chi = \mu = \left(\frac{\delta E}{\delta \rho(x)} \right)_V = V(x) + \frac{\delta F[\rho]}{\delta \rho(x)} \ . \qquad (2.77) $$

Equations 2.72, 2.76 and 2.77 are the governing equations of the chemical-physical reactivity processes, in terms of reactivity indices and density energetic functionals.

Further on, will be described the links between the sensitivity indices, at the local (space-dependent) and global (space averaged) levels, in a manner to allow an explicit implementation of the electronic densities.

The starting point generalizes the Euler-Lagrange equation 2.77 by introducing the density functional:

$$\Xi = \int \rho(x)V(x)dx + F[\rho] - \chi \int \rho(x)dx$$

$$= \int \rho(x)u(x)dx + F[\rho],$$

$$u(x) = V(x) + \chi .$$

(2.78)

Through searching its minimum,

$$\delta \Xi = 0$$

(2.79)

it leads with the equivalent Euler-Lagrange equation:

$$-u(x) = \frac{\delta F[\rho]}{\delta \rho(x)}.$$

(2.80)

The general fashion of equation 2.80 resides in its open electronic character by means of the combined influence potential 2.78.

Because of the general nature of the potential 2.78 in 2.80 it recommends as a sort of generalized electronegativity (local potential plus the global electronegativity) allowing therefore also that the associated generalized chemical hardness is likely to be introduced, by employing of both the nominator and the denominator of the generic 2.9 definition to the local behavior. This way, is generated the so called *chemical hardness kernel* associated to the system:

$$2\eta(x, x') = -\frac{\delta u(x)}{\delta \rho(x')}$$

$$= \frac{\delta^2 F}{\delta \rho(x) \delta \rho(x')}.$$

(2.81)

The definition 2.81 opens as well the possibility to introduce the inverse response of the electronic density under the variation of the generalized potential 2.78, namely *the chemical softness kernel* index:

$$s(x, x') = -\frac{\delta \rho(x)}{\delta u(x')}$$

(2.82)

assuming that such functional derivation exists and is real.

The relation between chemical hardness 2.81 and softness 2.82 kernels is not the direct inverse one but involving the delta Dirac identity,

$$\int \frac{\delta \rho(x)}{\delta u(x')} \frac{\delta u(x')}{\delta \rho(x'')} dx' = \delta(x''-x)$$

(2.83)

which produces the integral expression:

$$2 \int s(x, x') \eta(x', x'') dx' = \delta(x''-x).$$

(2.84)

The relations 2.81 and 2.82 reveals the most general reactivity indices for a given electronic system evolving in a given external potential, by reflecting the non-locality (through the two space non related variables) nature of the reactivity. The reverse way, the non-locality recuperating the locality, towards the global reactivity characterization, can be achieved by addressing the successive integrations (of averages) over the kernels and localities. Doing so, the *local chemical hardness,*

$$\eta(x) = \frac{1}{N} \int \eta(x, x') \rho(x') dx'$$

(2.85)

and *the local chemical softness*:

$$s(x) = \int s(x, x') dx'$$ (2.86)

are respectively introduced.

The remaining relations to be extracted among the indices 2.81-2.86 can be figured out appealing always the integration procedure. For instance, multiplying 2.84 with $\rho(x'')$ and then integrating upon x'', when equation 2.85 counts, there is obtained:

$$2\int s(x,x')\eta(x')dx' = \frac{1}{N}\rho(x)$$

(2.87)

from which, by next integration upon x, and accounting this time of the 2.50 and 2.86 relationships, follows the identity:

$$2\int s(x)\eta(x)dx = 1.$$

(2.88)

Now, one likes to combine these introduced kernel and local reactivity indices with the previous considered *DFT* exchange $[N, V]$ equations.

The strategy is to express the electronic density itself as an averaged response kernel reactivity index. Obvious, such representation have to follow the Hohenberg-Kohn theorems and their consequences that correlate the density with external applied potential. However, the general potential 2.78 will be here considerate. Successively, there are given out the relations:

$$d\rho(x) = \int \frac{\delta\,\rho(x)}{\delta\,u(x')}du(x')dx'$$

$$= -\int s(x,x')du(x')dx'$$

$$= -\int s(x,x')[dV(x')+d\chi]dx'$$

$$= -[\int s(x,x')dx']d\chi - \int s(x,x')dV(x')dx'.$$

(2.89)

Next, if there are also accounted of the equations 2.76 and 2.86, the equation 2.89 further becomes:

$$d\rho(x) = 2s(x)\eta\,dN + \int [-s(x,x'') + s(x)f(x'')]dV(x'')dx''. \quad (2.90)$$

The obtained equation 2.90 provides further meanings when also the alternative [N, V] differential expansion of the electronic density is assumed. The functional dependence $\rho = \rho[N, V]$ posses, however, a fully *DFT* stream by combining all required N- and V- uniquely dependencies. Appealing the Fukui function 2.59, the explicit [N, V] differential electronic density looks like:

$$d\rho(x) = f(x)dN + \int \left(\frac{\delta\rho(x)}{\delta V(x'')} \right)_N dV(x'')dx'' \ .$$

(2.91)

Now, because the equations 2.90 and 2.91 express the same electronic density reality, their identification will predict important reactivity indices' relationships.

By one-to-one correspondences of the first terms in the r.h.s. of equations 2.90 and 2.91, respectively, the next programmatic definitions of the local chemical softness are provided:

$$s(x) = \frac{f(x)}{2\eta}$$

$$= -\left(\frac{\partial\rho(x)}{\partial N} \right)_V \left(\frac{\partial N}{\partial \chi} \right)_V$$

$$= -\left(\frac{\partial\rho(x)}{\partial \chi} \right)_V$$

(2.92)

from which is possible to redefine again the Fukui function 2.59, this time in terms of the reactivity indices only dependence:

$$f(x) = \frac{s(x)}{S}$$

(2.93)

and where, the global chemical softness as the inverse of the doubly global chemical hardness was as well introduced:

$$S \equiv \frac{1}{2\eta} \ .$$

(2.94)

Therefore, the Fukui index 2.93 ultimately accounts of the contribution of the local softness to the global one.

However, this new Fukui definition, together with the identification between the second terms in the r.h.s. of equations 2.90 and 2.91, respectively, consecrates the role of the softness reactivity indices class (kernel, local, and the global ones) through the introduction of the *linear response function* for an electronic system, as:

$$\kappa(x, x'') \equiv \left(\frac{\delta \rho(x)}{\delta V(x'')} \right)_N$$

$$= -s(x, x'') + \frac{s(x)s(x'')}{S} .$$

$$(2.95)$$

It worth noting that the linear response function 2.95 is normalized to zero on behalf of the *DFT* normalization 2.50:

$$\int \kappa(x, x'') dx = \int \left(\frac{\delta \rho(x)}{\delta V(x'')} \right)_N dx$$

$$= \left(\frac{\partial N}{\partial V(x'')} \right)_N$$

$$= 0 .$$

$$(2.96)$$

The condition 2.96, in fact, will help a lot for the further procedure, allowing a practical density formulation for the chemical softnesses density class.

The next natural step is to try to express in a more explicitly manner the above introduced reactivity indices in terms of electronic density. Obvious, such an enterprise have to careful take into account the accepted conceptual quantum mechanical electronic description theorems.

An important theorem regards the above 2.96 linear response normalization. Another one is the invariance theorem, abstracted from the Hohenberg-Kohn ones, being specialized from the second identity of series 2.89, at the gradient level:

$$\nabla \rho(x) = - \int s(x, x') \nabla_{x'} u(x') dx' . \qquad (2.97)$$

The major advantage of using equation 2.97 as a working relation is that it succeeds to elude the knowledge of the exact Hohenberg-Kohn functional form $F_{HK}[\rho]$ in a way which maintains the integrity of the unique correspondence between ρ and V, according with basic *DFT* theorems. Last quantum mechanical theorem, here called to be fulfilled, addresses the Hellmann-Feynman theorem, [128] which tells that around the ground state is prescribed the equilibrium zero average of the exercised force $F(x)$ of applied external potential, through the series of identities:

$$0 = \int \rho (x) F(x) dx$$

$$= -\int \rho (x) \nabla V(x) dx$$

$$= -\int \rho (x) \nabla u(x) dx \ .$$

(2.98)

All these theorems become operative when one specific softness kernel picture is assumed. A practical, even as the *quasi* independent particle, model considers the separation between the local and non-local contributions in the softness kernel as: [129]

$$s(x, x') = L(x') \delta (x - x') + t(x) \rho (x')$$

(2.99)

The introduced functions, $L(x)$ and $t(x)$ of the local and non-local r.h.s terms in 2.99, respectively, have to be determined by applying, consecutively, the above translational invariance, Hellmann-Feynman and the linear response theorems. For instance, when submitting the expression 2.99 into the 2.97 one and by counting of the Hellmann-Feynman 2.98 prescription one simply gets:

$$\nabla \rho (x) = -L(x) \nabla u(x) - t(x) \int \rho (x') \nabla_{x'} u(x') dx'$$

$$= -L(x) \nabla u(x) \ .$$

(2.100)

Therefore, the local contribution $L(x)$ will take the equivalent forms:

$$L(x) = -\frac{\nabla \rho (x)}{\nabla u(x)}$$

$$= -\frac{\nabla \rho(x)}{\nabla V(x)}.$$

(2.101)

Up to now, the local and global softness can be respectively recuperated by the successive spatial integrations on the softness kernel 2.99, to give:

$$s(x) = L(x) + Nt(x),$$

$$S = \int [L(x) + Nt(x)] dx.$$

(2.102)

Next, the insertion of the softness kernel 2.99, together with the local and global softnesses 2.102 expressions, back into the linear response function 2.95 will lead it with the $t(x)$-dependency,

$$\kappa(x,x') = -L(x')\delta(x-x') - t(x)\rho(x') + \frac{[L(x) + Nt(x)][L(x') + Nt(x')]}{\int [L(x) + Nt(x)] dx}$$

(2.103)

which can be solved out by the further employing of the normalization condition 2.96. Applying so the condition 2.96 on 2.103, an integral equation is produced:

$$\int t(x) dx = \frac{N}{\rho(x)} t(x)$$

(2.104)

providing, in accord with the first *DFT* theorem - 2.50, the simple solution:

$$t(x) = \rho(x).$$

(2.105)

Finally, with expressions 2.101 and 2.105, the local and global softnesses 2.102 become the analytic density indices (in atomic units):

$$s(x) = -\frac{\nabla \rho(x)}{\nabla V(x)} + N\rho(x),$$

$$S = N^2 - \int \frac{\nabla \rho(x)}{\nabla V(x)} dx$$

(2.106)

with the help of which, the whole class of the reactivity descriptors, here presented, can be expressed within a consistently [ρ, V] DFT based context.

From now on, the analytical implementation is immediate, once an explicit expression of the atomic or molecular density is known.

A suitable analytical formalism for approaching the many-electronic densities calls the path integral picture. Its phenomenological base will be described in the next section.

2.2.5 The Path Integrals Quantum Formalism

A powerful insight in the quantum mechanical and quantum statistical description of many-body systems is given by the path-integral(s)-PI approach. This method, which was initiated by Feynman in his Princeton Thesis and the seminal '48 article, [130] provides an alternative formulation of the Schrödinger wave function formalism. In principle, the PI method is based on the quantum principle of superposition that allows to treat the transition amplitude between two states by the sum of amplitudes along all possible paths $x(\tau)$ connecting that states in a given time τ. Usually, the paths are parameterized by the imaginary time $i\hbar\beta$ being respectively, $\hbar$ the Planck's constant divided by 2π, and, β the notation for $1/k_BT$ where k_B is Boltzmann's constant. When the closed paths are considered, i.e. $x(0) = x(\hbar\beta)$, the PI partition function of the considerate system in its ground state can be extracted. [131] These are the general features that make the PI approach to be a complete quantum mechanical scheme, being able to treat both time-dependent as well as the equilibrium properties of a given system. In this framework worth to mention the last PI work of Feynman that becomes known as Feynman-Kleinert formalism. [24, 25, 132] That work proposes a sort of universal approximation method that has been tested in various fundamental and applied fields, from atoms and membranes, to stochastic and relativistic systems, being also in a continuous progress. [133] In the last decade, the PI methods have achieved a great interest by the chemical physical community providing applications that range from liquids and clusters up to the polymeric and biological systems. [134-151] For instance, the equilibrium properties of an excess electron in molten alkali halide and water, solvent and cluster, have been intensively investigated employing the PI pseudo-potentials to electron-solvent interaction model, using the isomorphism that maps the electron onto a classical polymer interacting with solvent molecules. [142] Electronic structure of ground states, tunneling splitting and canonical reaction rates have been successfully applied to study the equilibrium properties of excess electrons in clusters and classical fluids, e.g. the particle exchange in crystal ^{3}He or the electron transfer in aqueous Fe^{2+}/Fe^{3+} and $Co(NH_3)_6^{2+}/Co(NH_3)_6^{3+}$ by means of numerical *PI* methods. [151] Accurate and elaborate comparison between different numerical (discretized and Fourier) *PI* methods for quantum mechanical calculation of the free energies for the prototype molecular systems have been also recently performed. [149, 150] Biological systems have been as well investigated, in the light of the ability of the *PI* method to be coupled with the theory of barrier crossings, [135, 141] in revealing the spatial distribution and simulation of the real paths in the long-range electron transfer within protein environments, or for the charge separation in bacterial photosynthetic reaction centers. Iterative *PI* formulation over long time intervals providing a feasible simulation of complex quantum dissipative systems, have equally been recently reported. [147b]

The main *PI* idea had started from Feynman when he considered the partitioning of the lapse of time as having the following form:

$$[t_a, t = t_b] \rightarrow \Delta t = \frac{t_b - t_a}{n} \Rightarrow t_b = t_a + n\,\Delta t$$

$$(2.107)$$

which allows the rewriting the time evolution operator $\hat{U}(t_b, t_a)$ between the states (t_a, x_a), (t_b, x_b) by using the property of operator decomposition:

$$\left\langle x_b \middle| \hat{U}(t_b, t_a) \middle| x_a \right\rangle = \left\langle x_b \middle| \hat{U}(t_b, t_{b-1}) \hat{U}(t_{b-1}, t_{b-2}) ... \hat{U}(t_2, t_1) \hat{U}(t_1, t_a) \middle| x_a \right\rangle.$$

$$(2.108)$$

The above transition amplitude can be further rewritten, by inserting the spectral unit $1 = \int dx_j |x_j\rangle\langle x_j|$ between all the considered operatorial products of the temporal partition, and gives:

$$\left\langle x_b \middle| \hat{U}(t_b, t_a) \middle| x_a \right\rangle = \prod_{j=1}^{n-1} [\int dx_j] \prod_{j=1}^{n} \left\langle x_j \middle| \hat{U}(t_j, t_{j-1}) \middle| x_{j-1} \right\rangle.$$

$$(2.109)$$

Then, will be analyzed the amplitude of the partition transition, by considering the standard form of the characteristic Hamiltonian $\hat{H} = \hat{T}(p) + \hat{V}(x)$ of the studied electronic evolution. If the adopted temporal partition is sufficiently smooth so that $\Delta t \rightarrow 0$, one can uses the Hausdorff decomposition:

$$\exp[-i\frac{\hat{H}\Delta t}{\hbar}] \cong \exp[-i\frac{\hat{T}(p)\Delta t}{\hbar}]\exp[-i\frac{\hat{V}(x)\Delta t}{\hbar}].$$

(2.110)

With the help of 2.110 the partition of the transition amplitude can be successively transformed as:

$$\langle x_j|\hat{U}(t_j,t_{j-1})|x_{j-1}\rangle \cong \langle x_j|\exp[-i\frac{\hat{T}(p)\Delta t}{\hbar}]\exp[-i\frac{\hat{V}(x)\Delta t}{\hbar}]|x_{j-1}\rangle$$

$$= \int dx \int \frac{dp}{2\pi \hbar} \int \frac{dp_j}{2\pi \hbar}\langle x_j|\exp[-i\frac{\hat{V}(x)\Delta t}{\hbar}]|x\rangle\langle x|p_j\rangle\langle p_j|\exp[-i\frac{\hat{T}(p)\Delta t}{\hbar}]|p\rangle\langle p|x_{j-1}\rangle$$

$$= \int \frac{dp_j}{2\pi\hbar}\exp[-i\frac{V(x_j)\Delta t}{\hbar}]\exp[\frac{i}{\hbar}p_j(x_j - x_{j-1})]\exp[-i\frac{T(p_j)\Delta t}{\hbar}]$$

(2.111)

where have been used the spectral units on the coordinate and impulse space together with the following consecrated quantum operatorial relations: [25]

$$\langle x_j|\hat{V}(x)|x\rangle = V(x_j)\delta(x - x_j),$$

$$\langle p_j|\hat{T}(p)|p\rangle = 2\pi \hbar T(p_j)\delta(p - p_j),$$

$$\langle x|p_j\rangle = \exp[\frac{i}{\hbar}p_j x].$$

(2.112)

With the help of equation 2.111, the general transition amplitude 2.109 takes the form:

$$\langle x_b|\hat{U}(t_b,t_a)|x_a\rangle$$

$$= \prod_{j=1}^{n-1}[\int dx_j]\prod_{j=1}^{n}\int \frac{dp_j}{2\pi \hbar}\exp\left\{\frac{i}{\hbar}\sum_{j=1}^{n}[p_j(x_j - x_{j-1}) - \Delta t(T(p_j) + V(x_j))]\right\}$$

$$= \prod_{j=1}^{n-1} [\int dx_j] \prod_{j=1}^{n} \int \frac{dp_j}{2\pi\hbar} \exp\left\{ \frac{i}{\hbar} \sum_{j=1}^{n} \left[-\frac{\Delta t}{2m_0} \left(p_j - \frac{x_j - x_{j-1}}{\Delta t} m_0 \right)^2 + \frac{m_0}{2} \Delta t \left(\frac{x_j - x_{j-1}}{\Delta t} \right)^2 - \Delta t V(x_j) \right] \right\}$$

$$(2.113)$$

and, by using the Fresnel integral formula,

$$\int_{-\infty}^{\infty} dp_j \frac{1}{2\pi\hbar} \exp\left\{ -\frac{i}{\hbar} \frac{\Delta t}{2m_0} \left(p_j - \frac{x_j - x_{j-1}}{\Delta t} m_0 \right)^2 \right\} = \frac{1}{\sqrt{2\pi \hbar i \Delta t / m_0}}$$

$$(2.114)$$

the final transition amplitude results to be:

$$\langle x_b | \hat{U}(t_b, t_a) | x_a \rangle$$

$$= \frac{1}{\sqrt{2\pi \hbar i \Delta t / m_0}} \prod_{j=1}^{n-1} \left[\int \frac{dx_j}{\sqrt{2\pi \hbar i \Delta t / m_0}} \right] \exp\left\{ \frac{i \Delta t}{\hbar} \sum_{j=1}^{n} \left[\frac{m_0}{2} \left(\frac{x_j - x_{j-1}}{\Delta t} \right)^2 - V(x_j) \right] \right\}$$

$$\overset{\Delta t \to 0}{:=} \int_{x(t_a)=x_a}^{x(t_b)=x_b} Dx(t) \exp\left\{ \frac{i}{\hbar} \int_{t_a}^{t_b} dt\, L(x, \dot{x}, t) \right\}.$$

$$(2.115)$$

In 2.115 is recognized the exponent as the precisely classical action $A_{cl} = \int_{t_a}^{t_b} dt\, L(x, \dot{x}, t)$ for the Langranjean $L(x, \dot{x}, t) = m_0 \dot{x}^2 / 2 - V(x, t)$ associated to the electronic evolution under the influence of the $V(x)$ potential. The $\int Dx$ integral, represents the *path integral* which includes the integrations upon all the possible paths from x_a to x_b. The transition amplitudes from x_a to x_b in the lapse of time $[t_a, t_b]$ can be obtained totalizing all the possible paths and assigning to each of them a phase which corresponds to the classical associated action multiplied with the $i/\hbar$ factor. Next, using such a transition amplitude, one can rewrite the *total density operator*,

$$\hat{W}(t_b) = \hat{U}(t_b - t_a)\hat{W}(t_a)\hat{U}^*(t_b - t_a) \qquad (2.116)$$

under the forth ("+") and back ("–") paths decoupled form:

$$\hat{W}(x_b^+, x_b^-; t_b) = \left\langle x_b^+ \left| \hat{W}(t_b) \right| x_b^- \right\rangle$$

$$= \int dx_a^\pm \left\langle x_b^+ \left| \hat{U}(t_b, t_a) \right| x_a^+ \right\rangle \left\langle x_a^+ \left| \hat{W}(t_a) \right| x_a^- \right\rangle \left\langle x_a^- \left| \hat{U}^*(t_b, t_a) \right| x_b^- \right\rangle$$

$$= \int dx_a^\pm \int\limits_{x^+(t_a)=x_a^+}^{x^+(t_b)=x_b^+} Dx^+ \int\limits_{x^-(t_a)=x_b^-}^{x^-(t_b)=x_a^-} Dx^- \exp\left[\frac{i}{\hbar} \int\limits_{t_a}^{t_b} dt\, L(x^+, \dot{x}^+; t) \right] \left\langle x_a^+ \left| \hat{W}(t_a) \right| x_a^- \right\rangle$$

$$\times \exp\left[-\frac{i}{\hbar} \int\limits_{t_a}^{t_b} dt\, L(x^-, \dot{x}^-; t) \right].$$

(2.117)

    Further, one can decouple the associated Langranjean in 2.117 of an open electronic system upon its degrees of freedom into the quantum system $[Q, q]$, the reservoir with which is in contact $[R, r]$, and the coupling term $[Q\text{-}R]$:

$$L = L_Q(q) + L_R(r) + L_{Q-R}(q, r)$$

(2.118)

being both the variable x and the integrations limits in 2.117 as well decoupled on the (q, r) coordinates. Therefore, the effective electronic density matrix, of the quantum system $[Q, q]$, will be written as in 2.117, by cumulating all the effective electronics' ways back and forth between the extreme (t_a, q_a) and (t_b, q_b) events (points), respectively: [14]

$$\rho(q_b^+, q_b^-; t_b) = \int dr_b^\pm\, \hat{W}(q_b^+, r_b^+, q_b^-, r_b^-; t_b)$$

$$= \int dq_a^\pm dr_a^\pm dr_b^\pm \int Dq^+ Dq^- Dr^+ Dr^-\, \hat{W}(q_a^+, r_a^+, q_a^-, r_a^-; t_a)$$

$$\times \exp\left\{ \frac{i}{\hbar} \int\limits_{t_a}^{t_b} dt\left[L_Q(q^+, \dot{q}^+; t) - L_Q(q^-, \dot{q}^-; t) \right] \right\}$$

$$\times \exp\left\{\frac{i}{\hbar}\int_{t_a}^{t_b}dt\left[L_R(r^+,\dot{r}^+;t)-L_R(r^-,\dot{r}^-;t)\right]\right\}$$

$$\times \exp\left\{\frac{i}{\hbar}\int_{t_a}^{t_b}dt\left[L_{Q-R}(q^+,\dot{q}^+,r^+,\dot{r}^+;t)-L_{Q-R}(q^-,\dot{q}^-,r^-,\dot{r}^-;t)\right]\right\}.$$

(2.119)

If at the initial moment $(q_a^\pm, r_a^\pm, t_a)$, the density matrix is discretized such that containing the equilibrium information:

$$\hat{W}(q_a^+,r_a^+,q_a^-,r_a^-;t_a)=\rho(q_a^+,q_a^-;t_a)\hat{R}_{equilibrium}(r_a^+,r_a^-;t_a),$$

$$\hat{R}_{equilibrium}=\frac{\exp[-\beta\,\hat{H}_R]}{Tr\left\{\exp[-\beta\,\hat{H}_R]\right\}}$$

(2.120)

the effective quantum *PI* electronic density matrix 2.119 becomes:

$$\rho(q_b^+,q_b^-;t_b)=\int dq_a^\pm\int Dq^+Dq^-\,\rho(q_a^+,q_a^-;t_a)\overline{F}(q^+,q^-)$$

$$\times\exp\left\{\frac{i}{\hbar}\int_{t_a}^{t_b}dt\left[L_Q(q^+,\dot{q}^+;t)-L_Q(q^-,\dot{q}^-;t)\right]\right\}$$

(2.121)

and where, the so called Feynman-Vernon functional has been introduced: [14]

$$\overline{F}(q^+,q^-)=\int dr_a^\pm dr_b^\pm\int Dr^+Dr^-\exp\left\{\frac{i}{\hbar}\int_{t_a}^{t_b}dt\left[L_R(r^+,\dot{r}^+;t)-L_R(r^-,\dot{r}^-;t)\right]\right\}$$

$$\times\hat{R}_{equilibrium}(r_a^+,r_a^-;t_a)$$

$$\times \exp\left\{\frac{i}{\hbar}\int_{t_a}^{t_b}dt\left[L_{Q-R}(q^+,\dot{q}^+,r^+,\dot{r}^+;t)-L_{Q-R}(q^-,\dot{q}^-,r^-,\dot{r}^-;t)\right]\right\}.$$

(2.122)

The functional 2.122 contains interactions which are non-local in time and cover (theoretically) the entire evolution interval of the density matrix.

The consequence relies in the fact that the density matrix, within the path integral picture, contains all the memory effects, which means that its path integral corresponds to a non-markovian approach.

This fact runs against to the natural modeling of the chemical reactions, which are submitted to the principle according to, in evolution, the state of the electronic system depends on its immediately previous state and not on the entire accumulated memory. Therefore, within path integral reactivity modeling, the condition $t_b - t_a \to 0$, which recuperates the markovian character of the evolving structure of the studied electronic system, has to be implemented in an explicit way. [31] The advantage of this markovian modeling approach of the electronic reactivity, within the path integrals formalism, consists in the fact that the formulation is not a perturbative one, so that the solution (even temporary) of the effective quantum density (or of the density's matrix) is not restricted at the second order development of the perturbations in the $\hbar$ parameter, the way in which the mostly developments based on the Schrödinger equation are performed. [14] In this context, an essential step is made up by the passage from the quantum mechanics (QM) to the quantum statistics (QS), being the last quantum picture the most appropriate approach to treat the open systems. This passage is realized by considering the equivalence between the unitary evolution operator in the quantum mechanics:

$$\hat{U}(t_b,t_a)=\exp\left[-\frac{i}{\hbar}\hat{H}(t_b-t_a)\right]$$

(2.123)

and the effective density operator in the quantum statistics,

$$\hat{\rho}=\frac{1}{Z}\exp\left[-\frac{\hat{H}}{k_B T}\right]$$

(2.124)

based on the so-called Wick rotation: [31]

$$t_b - t_a = -i\frac{\hbar}{k_B T}=-i\hbar\beta\ .$$

(2.125)

The equivalence 2.125 will be marked here through the notation $t := -i\tau$. It can be observed therefore that, in the quantum statistics picture the real time is translated on the imaginary axis. The quantum statistic framework presents, however, also another advantage: the fact that the effective density is directly obtained like a statistic integral emphases on the quantum statistic distribution of the multi-electronic states.

There is useful to analyze how the Wick rotation 2.125 acts upon the transition density (or the transition amplitude).

For instance, for the often analyzed ω-harmonic external type potential the following Wick-based transformations are obtained:

$$\rho(x_b, t_b; x_a, t_a)_{QM}$$

$$= \int_{x(t_a)=x_a}^{x(t_b)=x_b} Dx(t) \exp\left\{\frac{i}{\hbar} \int_{t_a}^{t_b} dt \left[\frac{m_0}{2}\overset{\bullet}{x}^2(t) - \frac{m_0}{2}\omega^2 x(t)^2\right]\right\}$$

(2.126)

$$\downarrow \begin{cases} t := -i\tau \\[2mm] \dfrac{d}{dt} = \dfrac{d\tau}{dt}\dfrac{d}{d\tau} = i\dfrac{d}{d\tau} \end{cases}$$

$$\rho(x_b, \hbar\beta; x_a, 0)_{QS}$$

$$= \int_{x(0)=x_a}^{x(\hbar\beta)=x_b} Dx(\tau) \exp\left\{\frac{i}{\hbar} \int_0^{\hbar\beta} (-i)d\tau \left[\frac{m_0}{2}\left(i\frac{d}{d\tau}x(\tau)\right)^2 - \frac{m_0}{2}\omega^2 x(\tau)^2\right]\right\}$$

$$= \int_{x(0)=x_a}^{x(\hbar\beta)=x_b} Dx(\tau) \exp\left\{-\frac{1}{\hbar} \int_0^{\hbar\beta} d\tau \left[\frac{m_0}{2}\overset{\bullet}{x}^2(\tau) + \frac{m_0}{2}\omega^2 x(\tau)^2\right]\right\}.$$

(2.127)

Therefore, one will say that the classical harmonic Langranjean, $L(x, \overset{\bullet}{x}, t) = m_0 \overset{\bullet}{x}(t)^2/2 - m_0\omega^2 x(t)^2/2$, is so far transformed into the respective Euclidian Langranjean one, $L_e(x, \overset{\bullet}{x}, t) = m_0 \overset{\bullet}{x}(t)^2/2 + m_0\omega^2 x(t)^2/2$, analogous with the fact that the Euclidian metric has all the diagonal components +1.

The present study uses in next, for its computational part, the implementation of the electronic densities based on the evaluation quantum statistic picture of the path integrals, by means of the generalized Euclidian Langranjean $L_e(x,\dot{x},\tau) = m_0 \, \dot{x}(\tau)^2 / 2 + V(x(\tau))$. Afterwards, for the open systems analyzed in Chapter 3, the prototype harmonic and anharmonic molecular potentials are to be considered as the explicitly realization cases.

2.2.6 The Path Integral Feynman-Kleinert Formalism

As the *DFT* states, all the required main information about an electronic state are found in its external potential.

From a physical point of view this approach is based on the fact that the imposed external potential will determine the electronic density. [16, 105] The classical link between these two quantities is given by the Schrödinger equation, employing the wave function concept.

Alternatively, there is the *PI* formalism, that under its Feynman-Kleinert venture gives the electronic density through the electronic partition function (*Z*) and in which, instead of solving a differential equation – as the Schrödinger equation is, there is proposed to solve a parametrical integral. [15, 24]

Such a computation for the electronic density here follows, in terms of partition function, the same procedure as the Levy's constrained-search formalism does for the wave function, [18] see the Section 2.2.3.

Further on, will be exposed the construction of the quantum statistic electronic states for the effective densities of the electronic systems.

The calculus of the effective electronic density on the base of the path integrals starts, within the Feynman – Kleinert formalism, from the quantum statistic representation of the partition function: [24, 25]

$$Z = \oint_{x(0)=x(\hbar\beta)} Dx(\tau) \exp\left\{ -\frac{1}{\hbar} \int_0^{\hbar\beta} d\tau \left[m_0 \frac{\dot{x}(\tau)^2}{2} + V(x(\tau)) \right] \right\}$$

(2.128)

in which, the periodicity $x(0) = x(\hbar\beta)$ of the paths on the temporal (imaginary) axis is admitted, being the quantum statistical measure of integration in 2.128 the Wick-rotated quantum mechanical one from 2.115.

However, the normalization of the measure of the integral 2.128 is explicitly exposed in the Appendix, by the relations A.16 and A.43.

Since the Fourier decomposition of the periodical paths is considered,

$$x(\tau) = x_0 + \sum_{m=1}^{\infty}(x_m \exp[i\omega_m \tau] + c.c.)$$

(2.129)

being $\omega_m = 2\pi m/(\hbar\beta)$ the Matsubara frequencies, with m the integer number of indexation, the quantum statistic partition function 2.128 further becomes, see Appendix-A.43:

$$Z = \int_{-\infty}^{+\infty}\frac{dx_0}{\sqrt{2\pi\,\hbar^2\beta/m_0}}\left\{\prod_{m=1}^{\infty}\int_{-\infty}^{+\infty}\int_{-\infty}^{+\infty}\frac{d\,\mathrm{Re}\,x_m\,d\,\mathrm{Im}\,x_m}{\pi/(m_0\beta\omega_m^2)}\exp\left[\begin{array}{c}-\beta\,m_0\sum_{m=1}^{\infty}\omega_m^2|x_m|^2\\[4pt]-\dfrac{1}{\hbar}\int_0^{\hbar\beta}d\tau\,V(x(\tau))\end{array}\right]\right\}.$$

(2.130)

This form of the partition function has the advantage of including all the periodic paths, the quantum statistical ones, that characterize a given many – electronic ensemble, but having also the disadvantage of requiring the calculus of an infinite product of integrals. This is why the approximations are necessary, to enable the approach of the path integrals to become analytically applicable.

Firstly, will be formally rewritten the partition function 2.130 in a more compact form, which requires a single path integral only, governed by the formal *effective classical potential*, $V_{eff,cl}(x_0)$, instead of the entire Hamiltonian:

$$Z = \int_{-\infty}^{+\infty}\frac{dx_0}{\sqrt{2\pi\hbar^2\beta/m_0}}\exp[-\beta V_{eff,cl}(x_0)].$$

(2.131)

The x_0 variable represents the average position of all the possible quantum statistical paths on the imaginary temporal axis:

$$x_0 \equiv \overline{x} = \frac{1}{\hbar\beta}\int_0^{\hbar\beta}d\tau\,x(\tau)$$

(2.132)

and bears the name of the Feynman centroid. [31, 141]

In order to build up a formalism with a sufficient accuracy for the classical effective potential approximation, one has to consider a trial path integral, as a superposition of path integrals with harmonic potentials centered in different x_0 positions, each one of them having its own trial frequency, $\Omega^2(x_0)$.

Afterwards, the superposition and the respectively associated frequency will be chosen in the optimal way, so that the classical effective potential of the system should correspond to a quantum state as close as possible to that one approximated by the effective potential.

Consequently, the trial quantum statistic partition function (Z_1) successively becomes:

$$Z_1$$

$$= \oint Dx(\tau) \exp\left\{-\frac{1}{\hbar}\int_0^{\beta\hbar} d\tau \frac{m_0}{2}\left[\overset{\bullet}{x}^2 + \Omega^2(x_0)(x-x_0)^2\right]\right\} \exp[-\beta\, L_1(x_0)]$$

$$= \int_{-\infty}^{+\infty}\frac{dx_0}{\sqrt{2\pi\,\hbar^2\beta/m_0}}\prod_{m=1}^{\infty}\left[\int_{-\infty}^{+\infty}\int_{-\infty}^{+\infty}\frac{d\,\mathrm{Re}\,x_m\, d\,\mathrm{Im}\,x_m}{\pi/(m_0\beta\omega_m^2)}\right]$$

$$\times \exp\left\{-m_0\beta\sum_{m=1}^{\infty}[\omega_m^2 + \Omega^2(x_0)]|x_m|^2\right\}\exp[-\beta L_1(x_0)]$$

$$= \int_{-\infty}^{+\infty}\frac{dx_0}{\sqrt{2\pi\,\hbar^2\beta/m_0}}\frac{\hbar\beta\,\Omega(x_0)/2}{\sinh[\hbar\beta\,\Omega(x_0)/2]}\exp[-\beta\, L_1(x_0)]$$

$$\equiv \int_{-\infty}^{+\infty}\frac{dx_0}{\sqrt{2\pi\hbar^2\beta/m_0}}\exp[-\beta W_1(x_0)]$$

(2.133)

where the details of the integrations in the intermediate stages of expression 2.133 are enlighten in Section A.3 of the Appendix.

From the last line of the expression 2.133 the trial partition function appears now re-expressed in terms of the introduced trial potential W_1:

$$W_1(x_0) = \beta^{-1} \log \left\{ \frac{\sinh[\hbar \beta \, \Omega(x_0)/2]}{\hbar \beta \, \Omega(x_0)/2} \right\} + L_1(x_0)$$

(2.134)

and where, the functional $L_1(x_0)$, at its turn, will be later determined such that $W_1(x_0)$ to correspond to the optimal state, i.e. the closest one to the classical effective potential. In order to find the optimal state one will also rewrite the quantum statistic partition function 2.128 under the forms:

$$Z = \oint Dx \exp\left[-\frac{1}{\hbar} A_e \right]$$

$$= \oint Dx \exp\left[-\frac{1}{\hbar} A_{1e} \right] \exp\left[-\frac{1}{\hbar}(A_e - A_{1e}) \right]$$

$$= Z_1 \left\langle \exp\left[-\frac{1}{\hbar}(A_e - A_{1e}) \right] \right\rangle_1$$

(2.135)

in which the concerned bracket refers to the expected values calculated upon the general expression:

$$\langle O[x] \rangle_1 = Z_1^{-1} \oint Dx \exp\left[-\frac{1}{\hbar} A_{1e} \right] O[x]$$

(2.136)

being $A_{(1)e}[x,\dot{x}] = \int d\tau \, L_{(1)e}(x,\dot{x},\tau)$ the Euclidian version of the classical action, in the quantum statistic picture, see the end of the previous section.

Further on, by using the *Jensen-Peierls inequality*, [25]

$$\langle \exp[O] \rangle \geq \exp[\langle O \rangle]$$

(2.137)

based on the *convexity property* of the exponential function, and employing this variational principle on relation 2.135, there follows the inequality:

$$Z \geq Z_1 \exp\left[-\frac{1}{\hbar}\left\langle (A_e - A_{1e}) \right\rangle_1\right]$$

(2.138)

that in terms of free energy, $F = -\beta^{-1} \log Z$, transcribes as:

$$F \leq F_1 + \frac{1}{\beta\,\hbar}\left\langle A_e - A_{1e} \right\rangle_1.$$

(2.139)

At this point interferes the formalism's effectiveness: due to the fact that both Euclidian actions were built with the same kinetic term, $\int d\tau \left[m_0\, \dot{x}(\tau)^2 / 2 \right]$, from the difference in the expression 2.139 survives only the potential terms:

$$F \leq F_1 + \frac{1}{\hbar\beta}\int_0^{\hbar\beta} d\tau \left\langle V[x(\tau)] - \frac{m_0}{2}\Omega^2(x_0)\left(x(\tau) - x_0\right)^2 - L_1(x_0) \right\rangle_1.$$

(2.140)

The second step consists in evaluating the average $\langle\ \rangle_1$ in expression 2.140 throughout the Fourier development in the paths 2.129. On the base of 2.136, with the appropriate form of 2.133, the first term in 2.140 becomes:

$$\left\langle V(x(\tau)) \right\rangle_1 = Z_1^{-1} \int_{-\infty}^{+\infty} \frac{dx_0}{\sqrt{2\pi\,\hbar^2\beta / m_0}} \prod_{m=1}^{\infty}\left[\int_{-\infty}^{+\infty}\int_{-\infty}^{+\infty} \frac{d\,\mathrm{Re}\,x_m\, d\,\mathrm{Im}\,x_m}{\pi / (m_0\,\beta\omega_m^2)}\right]$$

$$\times \exp\left\{-\beta\, m_0 \sum_{m=1}^{\infty}[\omega_m^2 + \Omega^2(x_0)]|x_m|^2\right\}\exp[-\beta\, L_1(x_0)]$$

$$\times \int_{-\infty}^{+\infty}\frac{dk}{2\pi}\tilde{V}(k)\exp\left\{ik\left[x_0 + \sum_{m=1}^{\infty}x_m \exp(-i\omega_m\tau) + c.c.\right]\right\}.$$

(2.141)

Considering in the last expression the decomposition in perfect squares for the involved variables in the multiple integration, together with the notation:

$$a^2(x_0) = 2\frac{1}{m_0\beta}\sum_{m=1}^{\infty}\frac{1}{\omega_m^2 + \Omega^2(x_0)}$$

$$\underset{\text{(A.26)}}{\overset{\text{Appendix}}{=}} \frac{1}{m_0 \beta \, \Omega^2(x_0)} \left\{ \frac{\hbar \beta \, \Omega(x_0)}{2} \coth\left[\frac{\hbar \beta \Omega(x_0)}{2} \right] - 1 \right\}$$

$$(2.142)$$

the term 2.141 will successively be evaluated as:

$$\langle V(x(\tau)) \rangle_1 = Z_1^{-1} \int\limits_{-\infty}^{+\infty} \frac{dk}{2\pi} \tilde{V}(k) \int\limits_{-\infty}^{+\infty} \frac{dx_0}{\sqrt{2\pi \hbar^2 \beta / m_0}} \prod_{m=1}^{\infty} \left[\int\limits_{-\infty}^{+\infty} \int\limits_{-\infty}^{+\infty} \frac{d \, \mathrm{Re} \, x_m \, d \, \mathrm{Im} \, x_m}{\pi / (m_0 \beta \omega_m^2)} \right]$$

$$\times \exp\left\{ -m_0 \beta \sum_{m=1}^{\infty} [\omega_m^2 + \Omega^2(x_0)] \left[\left(\mathrm{Re} \, x_m - ik \frac{1/(m_0 \beta)}{\omega_m^2 + \Omega^2(x_0)} \cos\omega_m \tau \right)^2 \right. \right.$$

$$+ \left(\mathrm{Im} \, x_m - ik \frac{1/(m_0 \beta)}{\omega_m^2 + \Omega^2(x_0)} \sin\omega_m \tau \right)^2 \right] - \frac{a^2(x_0)}{2} k^2 + ikx_0 \bigg\} \exp[-\beta \, L_1(x_0)]$$

$$= Z_1^{-1} \int\limits_{-\infty}^{+\infty} \frac{dx_0}{\sqrt{2\pi \, \hbar^2 \beta / m_0}} \frac{\hbar \beta \, \Omega(x_0)/2}{\sinh[\hbar \beta \, \Omega(x_0)/2]}$$

$$\times \int\limits_{-\infty}^{+\infty} \frac{dk}{2\pi} \tilde{V}(k) \exp\left[ikx_0 - \frac{a^2(x_0)}{2} k^2 \right] \exp[-\beta \, L_1(x_0)]$$

$$(2.143)$$

where, for the performing way of the involved integrations the details and the results presented in the Appendix, A.31-A.38, have been used. Further on, by the insertion of the $\tilde{V}(k)$ integral Fourier representation,

$$\tilde{V}(k) = \int\limits_{-\infty}^{+\infty} dx'_0 \, V(x'_0) \exp[-ikx'_0]$$

$$(2.144)$$

and taking the integration over the wave vector k-space, again throughout the perfect square procedure, the expression 2.143 finally yields:

$$\langle V(x(\tau)) \rangle_1$$

$$= Z_1^{-1} \int_{-\infty}^{+\infty} \frac{dx_0}{\sqrt{2\pi \, \hbar^2 \beta / m_0}} \frac{\hbar\beta \, \Omega(x_0)/2}{\sinh[\hbar\beta \, \Omega(x_0)/2]} V_{a^2(x_0)}(x_0) \exp[-\beta \, L_1(x_0)] .$$

(2.145)

In expression 2.145, the new introduced potential:

$$V_{a^2(x_0)}(x_0) = \int_{-\infty}^{+\infty} \frac{dx'_0}{\sqrt{2\pi a^2(x_0)}} V(x'_0) \exp\left[-\frac{(x'_0 - x_0)^2}{2a^2(x_0)} \right]$$

(2.146)

appears from the original potential, the $V(x_0)$ one, by its expansion in the neighborhood of each effective event-point x_0, as a Gaussian package with the width $a^2(x_0)$. This modified-smeared out 2.146 potential takes in consideration all the quantum statistic fluctuations on the evolution of the considered electronic system. Analogous, one can develop the second term from the average in 2.140 so that the parametric dependence τ disappears, recovering the 2.145 result, but with the specialized smeared out potential 2.146 of the form:

$$(x - x_0)^2_{a^2(x_0)} = a^2(x_0) .$$

(2.147)

Since do not posses the τ parameter dependency, the last term from the average in 2.140 remains, throughout the smearing above 2.146 procedure, unchanged.

Cumulating all the obtained outlined averages back in 2.140 one gets:

$$\left\langle V(x(\tau)) - \frac{m_0}{2}\Omega^2(x_0)(x(\tau) - x_0)^2 - L_1(x_0) \right\rangle_1$$

$$= Z_1^{-1} \int_{-\infty}^{+\infty} \frac{dx_0}{\sqrt{2\pi \, \hbar^2 \beta / m_0}} \frac{\hbar\beta \, \Omega(x_0)/2}{\sinh[\hbar\beta \, \Omega(x_0)/2]}$$

$$\times \left[V_{a^2(x_0)}(x_0) - \frac{m_0}{2}\Omega^2(x_0)a^2(x_0) - L_1(x_0) \right] \exp[-\beta \, L_1(x_0)] .$$

(2.148)

Now, the parametric τ independency of average 2.148 permits the simple rewriting of the inequality 2.140 as the integrated out form:

$$F \le F_1 + \left\langle V_{a^2(x_0)}(x_0) - \frac{m_0}{2}\Omega^2(x_0)a^2(x_0) - L_1(x_0) \right\rangle_1 .$$

(2.149)

Analyzing the last inequality one can observe directly that the best variational choice, such that the two above free energies approach each other as close as possible, corresponds to the specialization of the function $L_1(x_0)$ to become:

$$L_1(x_0) = V_{a^2(x_0)}(x_0) - \frac{m_0}{2}\Omega^2(x_0)a^2(x_0) .$$

(2.150)

With the optimal choice given in 2.150, the free energy F is directly bounded by the free energy F_1,

$$F \le F_1$$

(2.151)

that also prescribes the similar inequality in terms of the partition functions, for the exact and approximate quantum statistical states:

$$Z \ge Z_1 .$$

(2.152)

Using the formulations 2.131 and 2.133, the relation 2.152 transcribes as:

$$\int_{-\infty}^{+\infty} \frac{dx_0}{\sqrt{2\pi \, \hbar^2 \beta / m_0}} \exp[-\beta \, V_{eff,cl}(x_0)] \ge \int_{-\infty}^{+\infty} \frac{dx_0}{\sqrt{2\pi \, \hbar^2 \beta / m_0}} \exp[-\beta \, W_1(x_0)]$$

(2.153)

that further prescribes the upper boundary potential W_1 for the classical effective one,

$$V_{eff,cl}(x_0) \le W_1(x_0)$$

(2.154)

and where, now, $W_1(x_0)$ takes the analytical expression:

$$W_1(x_0) = \frac{1}{\beta}\log\left\{\frac{\sinh[\hbar\beta\,\Omega(x_0)/2]}{\hbar\beta\,\Omega(x_0)/2}\right\} + V_{a^2(x_0)}(x_0) - \frac{m_0}{2}\Omega^2(x_0)a^2(x_0)$$

$$(2.155)$$

being this formula obtained by combination of the relations 2.134 and 2.150.

The exposed Feynman-Kleinert variational approach indicates that the *PI* formalism gives to partition function the same physical meaning that the wave function has in the Levy constrained-search *DFT* formalism.

For instance, the conditions 2.152 and 2.154 can be seen as the counterparts in the first step of the Levy's *DFT* algorithm in terms of partition function: from the all possible effective potentials, $V_{eff,cl}$, that give to partition function Z (see 2.131) the same meaning as the full Hamiltonian partition function Z (see 2.128), there are selected that ones, W_1, that closely approximates the entire Hamiltonian through the partition function Z_1 (see 2.133). The second step of the Levy's variational *DFT* approach can be recuperated at the *PI* level by performing the further minimization of the potential $W_1(x_0)$ respecting with the trial frequency $\Omega^2(x_0)$:

$$\frac{dW_1(x_0)}{d\Omega^2(x_0)} = \frac{\partial W_1(x_0)}{\partial\Omega^2(x_0)} + \frac{\partial W_1(x_0)}{\partial a^2(x_0)}\bigg|_{\Omega(x_0)}\frac{\partial a^2(x_0)}{\partial\Omega^2(x_0)}$$

$$= 0.$$

$$(2.156)$$

The first term from equation 2.156 can be further evaluated based on the equivalence $2(\partial/\partial\Omega^2)\bullet = (1/\Omega)(\partial/\partial\Omega)\bullet$ and leads with the result:

$$\frac{\partial W_1(x_0)}{\partial\Omega^2(x_0)} = \frac{m_0}{2}\left\{\frac{k_BT}{m_0\Omega^2(x_0)}\left[\frac{\hbar\Omega(x_0)}{2k_BT}\coth\left(\frac{\hbar\Omega(x_0)}{2k_BT}\right)-1\right]-a^2(x_0)\right\}$$

$$= 0$$

$$(2.157)$$

by recalling also the relations 2.155 and 2.142.

Therefore, from the condition 2.156 just remains the equation:

$$\frac{\partial W_1(x_0)}{\partial a^2(x_0)} = 0 \qquad\qquad (2.158)$$

from which follows the relation:

$$\Omega^2(x_0) = \frac{2}{m_0} \frac{\partial V_{a^2(x_0)}(x_0)}{\partial a^2(x_0)}$$

$$= \frac{1}{m_0} \left[\frac{\partial^2}{\partial x_0^2} V_{a^2(x_0)}(x_0) \right]_{a^2 = a^2(x_0)}$$

(2.159)

when the potential form 2.155 counts in equation 2.158.

However, the last equality in 2.159 comes from the derivation performed on the equivalent Fourier expression of the smeared out potential 2.146, namely with form:

$$V_{a^2(x_0)}(x_0) = \int\limits_{-\infty}^{+\infty} \frac{dk}{2\pi} \tilde{V}(k) \exp\left(ikx_0 - \frac{a^2(x_0)}{2} k^2 \right).$$

(2.160)

Resuming, the Feynman-Kleinert variational algorithm in path integrals provides the calculation of the effective electronic density by constructing the constraint-searched partition function picture as the phenomenological equivalence for what the Levy's constraint-search formalism describes for wave function. The main steps of this *PI* algorithm can be summarized as follows.

(i)   The partition function is given as a simple integral:

$$Z_1 = \int\limits_{-\infty}^{+\infty} \frac{dx_0}{\sqrt{2\pi \hbar^2 \beta / m_0}} \exp[-\beta W_1(x_0)].$$

(2.161)

(ii)   The introduced potential $W_1(x_0)$ satisfies the optimal form:

$$W_1(x_0) = \frac{1}{\beta} \log\left\{ \frac{\sinh[\hbar\beta \, \Omega(x_0)/2]}{\hbar\beta \, \Omega(x_0)/2} \right\} + V_{a^2(x_0)}(x_0) - \frac{m_0}{2} \Omega^2(x_0)a^2(x_0).$$

(2.162)

(iii)   The smeared out external potential looks like:

$$V_{a^2(x_0)}(x_0) = \int\limits_{-\infty}^{+\infty} \frac{dx'_0}{\sqrt{2\pi a^2(x_0)}} V(x'_0) \exp\left[-\frac{(x'_0 - x_0)^2}{2a^2(x_0)}\right].$$

(2.163)

(iv)   The introduced $a^2(x_0)$ and $\Omega^2(x_0)$ parameters fulfill the relations:

$$a^2(x_0) = \frac{1}{m_0\beta\,\Omega^2(x_0)}\left\{\frac{\hbar\beta\,\Omega(x_0)}{2}\coth\left[\frac{\hbar\beta\Omega(x_0)}{2}\right] - 1\right\},$$

$$\Omega^2(x_0) = \frac{2}{m_0}\frac{\partial V_{a^2(x_0)}(x_0)}{\partial a^2(x_0)}.$$

(2.164)

A concrete atomic specialization of this general PI model, suited for the electronic DFT framework, will be at the end of the this chapter presented.

2.3 TOWARDS A DENSITY FUNCTIONAL ELECTRONEGATIVITY THEORY

In the present section there will be introduced the density functionals with a reactivity role on the many-electronic systems.

This way, is assured the very general characterization of the chemical driving potency (on physical bases) for the analyzed electronic system, as long as the concerned system is involved in coordination, interaction, exchange and transformation processes.

2.3.1 The Chemical Action Principle

Since the exact analytical form of the Hohenberg-Kohn functional in DFT total energy decomposition 2.52 is still not known, it makes sense to separate and nominate the external action term 2.53 as the chemical action:

$$C_A[\rho] = \int \rho(x)V(x)dx.$$

However, the ground upon which the quantity 2.53 is an action is a direct consequence of the fundamental DFT theorems, presented in the Section 2.2.3, and is based on the universal nature of $F_{HK}[\rho]$, which does not depend on of the external potential $V(x)$.

To justify the last assertion, let's note that the stationary principle 2.54 provides the variational principle of the chemical action:

$$C_A[\bar{\rho}] \geq C_A[\rho] \Leftrightarrow \delta\, C_A[\rho] = 0 \,.$$

(2.165)

This way, because the exact analytical form of the functional $F_{HK}[\rho]$ is not accessible in 2.54, the chemical action $C_A[\rho]$ assumes, through the principle 2.165, a fundamental role in the structure and modification of the many-electronic ground states within *DFT* framework.

Nevertheless, the principle 2.165 supports also an alternative derivation, upon the differential energy equation, showing so far its general validity. To start, by remembering that $\chi = -\mu$, the equation 2.72 of the total energy can just be rewritten as:

$$dE - \mu dN = \int \rho(x) dV(x) dx \,.$$

(2.166)

Next, by the integration of the equation 2.166 followed by its functional differentiation,

$$\delta\left\{\int[dE - \mu\, dN]\right\} = \delta\left\{\int\left[\int\int\rho(x) dV(x) dx\right]\right\}$$

(2.167)

and by counting on the ground state property [18]:

$$\mu = CONSTANT$$

(2.168)

there is provided the identity:

$$\delta\left\{E[\rho] - \mu N[\rho]\right\} = \delta C_A \,.$$

(2.169)

Now, the interpretation of C_A is a more direct one: as far as the left hand side of equation 2.169 establishes the *stationary principle* in Density Functional Theory-*DFT*, see equation 2.55:

$$\delta\{E[\rho] - \mu N[\rho]\} = 0$$

while the right hand side formally recovers the *variational principle* for the system's action:

$$\delta\, C_A = 0 \quad .$$

(2.170)

In this way, C_A assumes the *chemical action* concept.

However, worth to note that it was Mel Levy the first one who, [152] using the universal properties of the Hohenberg-Kohn functional, had arrived to the above chemical action concept, available both for an arbitrarily large M-set of non-interacting as well as of interacting Hamiltonians, through the minimization of the G his functional:

$$G_{1,2,...,M}^{\alpha,\beta,...,\omega} = \int dx \Big[\rho_\alpha(x)V_1(x) + \rho_\beta(x)V_2(x) + ... + \rho_\omega(x)V_M(x) \Big].$$

(2.171)

The minimum of functional 2.171, states in fact a realization of the chemical action principle 2.170, and optimizes the ordering pairs of the densities with the associate external potentials.

This one-to-one optimization fully recovers the first Hohenberg-Kohn *DFT* prescription.

However, in deriving the chemical action principle, equations 2.165 and 2.170, the second Hohenberg-Kohn theorem was as well used, since the stationary principle 2.55 has been involved.

Therefore, the chemical action concept and its application fills the practical realization of the combined Hohenberg-Kohn theorems. In this its property resides the immense potentiality for the practicing of this concept to derive and control the density functionals. An illustration will be later presented regarding the *DFT* Mulliken electronegativity formulation.

2.3.2 The Chemical Field Concept

Before passing to the direct electronegativity analytical analysis, worth to introduce another reactivity index related with the chemical action one.

Performing the formal integration of the equation 2.72 there is obtained:

$$\int \chi dN = -E + \int \rho(x)V(x)dx \quad .$$

(2.172)

The equation 2.172 has an identical form with a Legendre transformation of the total energy E, which, from a functional dependency on the external potential $V(x)$, transforms itself into a density functional dependency, through the functional relation 2.73.

However, the relation 2.172 will be used for introducing a new descriptor, which will be directly related to the form of the quantum statistical field transformations. From these reasons, the deduced descriptor will next be called like the *chemical field*.

In order to arrive to the chemical field, the electronic partition function will be considered in the presence of a supplementary potential, the charge current with temporal dependency $j(\tau)$, as: [25]

$$Z[j] = \oint Dx(\tau) \exp\left\{ -\frac{1}{\hbar} \int_0^{\hbar\beta} d\tau \left[m_0 \frac{\dot{x}(\tau)^2}{2} + V(x(\tau)) - j(\tau)x(\tau) \right] \right\}$$

(2.173)

where $\oint Dx(\tau)$, represents the path integral 2.128 and includes the integration upon all closed paths between two spatial coordinates x_a and x_b.

Once a constant external source $j(\tau) \equiv j$ is concerned, the relation 2.173 turns from a functional to a function that depends only upon the j source:

$$Z(j) = \oint Dx(\tau) \exp\left\{ -\frac{1}{\hbar} \int_0^{\hbar\beta} d\tau \left[m_0 \frac{\dot{x}(\tau)^2}{2} + V(x(\tau)) \right] + \beta \, jx_0 \right\}.$$

(2.174)

Following the effective classical potential formalism, as in previous section was described for the Feynman-Kleinert approach, the path integral 2.174 can be reduced to a single integral.

So, within the effective classical potential $V_{eff,cl}$ approach, the partition function $Z(j)$ can be written like an unique integral upon the x_0 average as:

$$Z(j) = \int_{-\infty}^{+\infty} \frac{dx_0}{\sqrt{2\pi\beta \, \hbar^2 / m_0}} \exp\left\{ -\beta \left[V_{eff,cl}(x_0) - jx_0 \right] \right\}.$$

(2.175)

Once the partition functional 2.173 and the partition function 2.175 are considered, the following filed constructions can be carried out, respectively:

$$W[j] \equiv \log Z[j],$$

$$W(j) = \log Z(j).$$

(2.176)

Since one starts from these expressions, there can be correspondingly defined *the system's effective actions*, throughout the Legendre transformations, namely:

$$\Gamma[X] \equiv -W[j] + \int_0^{\hbar\beta} X(\tau)\, j(\tau)\, d\tau\,,$$

(2.177)

$$V_{eff,cl}(X) = -\frac{1}{\hbar\beta} W(j) + Xj\,.$$

(2.178)

In relation 2.177 the variable $X(\tau)$ states as the field average of the positions, and is functionally related with the source $j(\tau)$ by the relation:

$$X(\tau) \equiv \frac{\delta}{\delta\, j(\tau)} W[j]$$

(2.179)

whereas, in relation 2.178, the effective potential was introduced as the density of the effective action at the constant average position $X \equiv X(\tau)$, namely:

$$V_{eff,cl}(X) \equiv \frac{1}{\hbar\beta} \Gamma[X]\Big|_{X(\tau)\equiv X}\,.$$

(2.180)

Once the relations 2.177 with 2.179 are compared with the relations 2.172 and 2.73, respectively, the formal next correspondences can be reached out:

$$\Gamma[X] \leftrightarrow \int \chi\, dN,$$

$$W[j] \leftrightarrow E,$$

$$j(\tau) \leftrightarrow V(x).$$

(2.181)

The correspondences 2.181 show how the integrated electronegativity of the electronic system corresponds to the effective action in field theories. This observation can be further employed in a more deeply way and shall certain enrich the electronegativity phenomenology. However, by considering also the correspondences 2.178 with the equation 2.172, likewise the relations 2.181 were obtained, new fruitful associations are released:

$$\int \chi \, dN + E \leftrightarrow \hbar \beta j X$$

$$\Leftrightarrow C_A \leftrightarrow \hbar \beta j X .$$

(2.182)

Since above the same proportionality factor is about, by combining the last forms of 2.181 and 2.182, the so called *chemical field* is getting out:

$$X \equiv \omega_C = \frac{C_A}{\hbar \beta \, V(x)}.$$

(2.183)

The chemical field 2.183 combines both the density and the external potential local dependence through the intimate relation with chemical action 2.53. Therefore, the chemical field is correlated with the chemical action of the electronic state in a given external potential, in which the proportionality factor β corresponds to the "thermal bath" within the electronic system is immersed. Likewise, one remarks the quantum character of the chemical field's descriptor through the presence of the reduced Planck constant $\hbar$. A possible employment of the chemical field will be soon dedicated to the rate reaction calculations. More, from relation 2.183 one can also define *the chemical field's period:*

$$T_C = \frac{2\pi}{\omega_C}.$$

(2.184)

In these conditions, the chemical field has an oscillatory structure and is characterized by an *"amplitude"* and a *"period"*. An evaluation of the chemical field 2.183 and of the correspondent period 2.184 will be presented in Chapter 3 for a generalized anharmonic potential (in atomic units:

$V_{|g\rangle} = x^2/2 + gx^4/4$), characteristic for the open (coupled) molecular electronic systems. In such a case, the chemical field represents the response of the molecular electronic system for the simultaneous presence of a nuclear and/or electromagnetic field (g) and a thermal exchange (β).

If one considers only the external influence of an electromagnetic field, the chemical field would register the resonance frequency of this coupling, while, in the exclusive presence of the thermal exchange the chemical field would register a frequency characteristic to the thermal transition state.

For the electromagnetic-thermal field treatment, the chemical field would therefore express the combined throb of the two above superimposed influences.

2.3.3 New Electronegativity Density Functional

For a chemical-physical process, the equation of the electronegativity variation of an N-electronic system subjected to an external potential $V(x)$, is given by equation 2.76.

One can remark the fact that the electronegativity of an electronic state depends on the associated chemical hardness η (the component part which inhibits the reactivity) and on the spatial integrated frontier function $f(x)$ (the component part which promotes the reactivity).

One likes to integrate the equation 2.76 in order to find the electronegativity functional associated with the electronic state $|\rho_V\rangle$.

With the help of expressions 2.94 and 2.93, seen for the global hardness and the Fukui function definitions, respectively, the equation 2.76 of the electronegativity can be integrated to give:

$$\chi(N) = -\int_0^N \frac{1}{S} dN - \frac{1}{S} \int_{-\infty}^{+\infty} s(x)V(x)dx .$$

(2.185)

The integrals in 2.185 can be carried out once an analytical realization of the local and global softness $s(x)$ and S, respectively, is formulated. This can be achieved assuming a *quasi* independent-particle model within the *DFT* framework that provides the softness kernel $s(x, x')$ with expression 2.99 and the further local $s(x)$ and global S softnesses with their 2.106 expressions.

Consequently, the electronegativity states as a density functional, for which the analytical expression is determined by introducing the relations 2.106 in the equation 2.185 and performing out the integration.

This way, there follow the identities:

$$\chi(N) = -\int_0^N \frac{1}{\int_{-\infty}^{+\infty} L(x)dx + N^2} dN - \frac{\int_{-\infty}^{+\infty} L(x)V(x)dx + N\int_{-\infty}^{+\infty} \rho(x)V(x)dx}{\int_{-\infty}^{+\infty} L(x)dx + N^2}$$

$$= -\frac{1}{\sqrt{a}}\arctan\left(\frac{N}{\sqrt{a}}\right) - \frac{b}{a + N^2} - NC_A \frac{1}{a + N^2}$$

(2.186)

where, in order to simplify the above expression, apart of the recognized chemical action functional 2.53, additional definitions were introduced:

$$a = \int_{-\infty}^{+\infty} L(x)dx ,$$

$$b = \int_{-\infty}^{+\infty} L(x)V(x)dx$$

(2.187)

with $L(x)$ fixed by expression 2.101.

Worth to remark the fact that the electronegativity's "initial" value, corresponding to the free electronic system, was assumed to be zero. This is based on the fact that, since *electro-negativity* measures the potency of attracting electrons, in the absence of the nuclear or external applied filed the free electronic system does not attract additional electrons while the repulsive character is the dominant one. Therefore, the zero initial electronegativity value in the limit $V(x) \to 0$ is well justified. Indeed, even analytically, once the limit $V(x) \to 0$ applies on the expression 2.186 it produces the direct result:

$$\lim_{V(x) \to 0} \chi(N) = 0.$$

(2.188)

Next, worth to check the binding reliability of the electronegativity 2.186. There are well known the failures of the Thomas-Fermi statistical model for chemical potential (negative electronegativity), because predicts the zero electronegativity behavior in the large spatial limit, so to speak, in the zero electronic density limit $\rho(x \to \infty) \to 0$. This means that the Thomas-Fermi model precludes the molecular binding, [153] in contrast with the essence of what the electronegativity concept, that just this ability of forming composed electronic systems, likes to indicate beyond of the isolated quantum systems. Therefore, a suitable check of any proposed electronegativity formulation have to treat the zero density limit. However, one adopts a proper hierarchy of this limit, based on the fundamental *DFT* relation 2.50, and of its implications. This way, the order to apply this test prescribes the successive limits: $N \to 0$: $\rho \to 0$: $\nabla\rho \to 0$: ....

Proceeding in such with the electronegativity 2.186, one gets:

$$\lim_{(N,(\nabla\rho)) \to 0} \chi(N) = -\lim_{\nabla\rho \to 0} \frac{b}{a}$$

$$= -\frac{\int \dfrac{V(x)}{\nabla V(x)} dx}{\int \dfrac{dx}{\nabla V(x)}} \neq 0$$

$$(2.189)$$

a result that obvious is non-zero being potential dependent – a fact that closely allows the binding processes.

Concluding, the present electronegativity density functional, deduced here as the relation 2.186, displays an expression with a high level of generality, allowing its evaluation and interpretation for different electronic systems and cases.

2.3.4 New Rate Reaction Functional

By linking in a more deeply manner the physical and chemical many-body approaches, for instance by combining the *DFT* with Fokker-Planck (*FP*) path integral (*PI*) methods, [6, 7, 28-31, 39] new reactivity indices can be further inferred, in a most general manner.

If the key of such a link regards the *chemical potential,* μ (one of the fundamental physical quantities of the open systems), and its negative version that identifies with the *electronegativity,* $\chi = -\mu$ (the fundamental chemical counterpart quantity which reflects the reactivity ability of the concerned systems), the *rate reaction index* (*RR*) is springing out, as next showed.

Since the chemical potential is seen like the minus electronegativity, as generally prescribed in the conceptual quantum chemical approach-see equation 2.6, its analytical form (in atomic units) becomes:

$$\mu = \frac{1}{\sqrt{a}} \arctan\left(\frac{N}{\sqrt{a}}\right) + \frac{b}{a + N^2} + NC_A \frac{1}{a + N^2} \qquad (2.190)$$

where, the already deduced electronegativity formulation 2.186 was employed.

Next, eliminating the chemical action C_A between the equations 2.190 and 2.183, of the chemical potential and that of the chemical field, respectively, there is provided the actual rate reaction functional expression, as:

$$\omega_C \equiv \Gamma[\rho] = \frac{\left[\mu - \dfrac{1}{\sqrt{a}} \arctan\left(\dfrac{N}{\sqrt{a}}\right)\right](N^2 + a) - b}{\bar{t} N C_A},$$

$$\bar{t} = \hbar\beta$$

(2.191)

where, instead of preserving the external potential denominator dependence, the chemical action was recuperated, on the ground of their spatial average equivalence $<V(x)>=C_A$, accordingly with the basic definition 2.53.

Through the 2.191 dependence, the *rate reaction* assumes the meaning of frequency with which the considered system, having the chemical action $C_A[\rho]$, can surpass the barrier $V(x)$ within the temporal interval Δt that becomes so far *the mean first passage time* $\bar{t}$. [7-9, 28, 31] Therefore, by the significance of the 2.191 functional, its rate reaction phenomenology is assured.

However, in order to complete the present rate reaction formulation, an alternatively implementation of the chemical potential μ have to be considered.

For the physical expression of the chemical potential μ one will start to write it as the Gibbs potential G *per* particle:

$$\mu = \frac{G}{N}.$$

(2.192)

Maintaining the general character of the actual approach, the Gibbs potential can be further expressed as the sum of the free energy F and of the applied work function:

$$l = -\int \nabla V(x) dx.$$

(2.193)

Furthermore, considering the statistical picture of the free energy in terms of partition function Z and of the inverse thermal energy β, $F = \beta^{-1}\ln Z$, the resulting chemical potential looks like:

$$\mu = \frac{1}{N\beta}\ln Z - \frac{1}{N}\int \nabla V(x)dx \ .$$

(2.194)

The practical evaluations of the chemical potential 2.194 require the knowledge of the Z.

At this point, there can be appealed the *PI* picture of the partition function that, within *FP* approach, it replaces with the *conditional probability density*:

$$Z \rightarrow P(x_b, t_b; x_a, t_a)$$

(2.195)

by which, the event x_b is realized at the moment t_b once the event x_a is realized at the moment t_a.

Therefore, the chemical potential 2.194 of N identical particles, each of mass m_0, rewrites as:

$$\mu_{FP} = \frac{m_0\kappa}{N}\left\{\frac{1}{D}\ln[P(x_b, t_b; x_a, t_a)] + \int K(x)dx\right\}$$

(2.196)

where the diffusion D and the drift $K(x)$ coefficients, together with the friction constant κ, have appeared by performing in equation 2.196, respectively both the Einstein relation,

$$\frac{1}{\beta} = \frac{m_0\kappa}{D}$$

(2.197)

and the drift equation one: [27]

$$\nabla V(x) = -m_0\kappa \ K(x) \ .$$

(2.198)

At its turn, the conditional probability density $P(x_b, t_b; x_a, t_a)$ satisfies the *FP* equation, [28, 39]

$$\frac{\partial}{\partial t_b}P(x_b, t_b; x_a, t_a) = -\frac{\partial}{\partial x_b}[K(x_b)P(x_b, t_b; x_a, t_a)] + D\frac{\partial^2}{\partial x_b^2}P(x_b, t_b; x_a, t_a) \quad (2.199)$$

with the initial delta-Dirac condition,

$$P(x_b, t_a; x_a, t_a) = \delta(x_b - x_a)$$

(2.200)

of which solution has the *PI* representation: [39]

$$P(x_b, t_b; x_a, t_a)$$

$$= \int_{x(t_a)=x_a}^{x(t_b)=x_b} Dx(t) \exp\left\{ -\frac{1}{4D} \int_{t_a}^{t_b} dt \left[\dot{x}(t) - K(x(t)) \right]^2 - \frac{1}{2} \int_{t_a}^{t_b} dt K'(x(t)) \right\}$$

(2.201)

and where, all the paths $x(t)$ contribute in connecting the space-time points (x_a, t_a) and (x_b, t_b).

By assuming the initial density $\rho(x_a, t_a)$ of the system such that to satisfy the *DFT*'s first *HK* theorem, see equation 2.50, the actual density $\rho(x_b, t_b)$, as well the mean first passage time $\bar{t}$ in 2.191, can be expressed in terms of the conditional probability density 2.201, once a specific potential barrier and its proper drift are given, according with equation 2.198.

Finally, the chemical potential 2.196 and, therefore, also the *RR* expression 2.191, are computed within the path-integral *FP* approach.

Thus, through considering of the association of the physical (*FP* path integral) and chemical (*DFT*) expressions for the chemical potential and electronegativity, respectively, a new rate reaction (*RR*) formulation and its functional realization can be figured out.

This way, the new form of *RR* index assures the most general chemical-physical treatment aiming the characterization of the exchange, coordination and reactivity behavior for the many-electronic open systems. [5, 6]

2.3.5 Electronegativity Related Energetic Functionals

Once the electronegativity 2.186 is known analytically, it permits the total energy derivation throughout the equation 2.72. The integration of this equation will assume the initial total energy condition as the zero electronic energy in the absence of electrons or in the long range limit:

$$E(N \to 0 : \rho \to 0 : \nabla\rho \to 0 : ...) \to 0.$$

(2.202)

Plugging the electronegativity 2.186 in equation 2.72, and by integrating it under the condition 2.202, one formally yields:

$$E = -\int_0^N \chi(N)dN + C_A.$$

(2.203)

By comparing the expression 2.203 with the fundamental *DFT* total energy decomposition 2.51 there is performed the direct one-to-one correspondence, from which the Hohenberg-Kohn functional springs out to be as well electronegativity related, as:

$$F_{HK} = -\int_0^N \chi(N)dN.$$

(2.204)

Although by solving the integrals in expressions 2.203 and 2.04, when electronegativity 2.186 is considered, the obtained functionals give the energetic terms, they will be called as *partial* functionals for the reason that the both background equations, 2.76 and 2.72, were truncated to the first order in charge and external potential variation for changing from one ground state to another, respectively.

Nevertheless, the present results display a high level of generality, based on the total differential character of electronegativity and of the total energy of the concerned electronic system, a fact fully satisfied throughout the used above equations.

However, in order to don't produce a linguistically contradiction for what means the *partial* total energy, the adopted energetic terminology for the present functionals will address the *partial Hohenberg-Kohn functional* (F_{HK}^P) and the *reduced total energy functional* (E^R), respectively.

Next on, their analytical developments proceed.

By submitting the electronegativity 2.186 in the expression 2.204 the partial *HK* functional takes, after the direct integration, the analytically form:

$$F_{HK}^P = \frac{N+b}{\sqrt{a}}\arctan\left(\frac{N}{\sqrt{a}}\right) + \frac{C_A-1}{2}\ln\left(\frac{a+N^2}{a}\right)$$

$$\equiv F_A + F_B.$$

(2.205)

Obvious, with 2.205 in 2.203, the reduced total energy assumes the form:

$$E^R = F_A + F_B + C_A$$

$$= -T$$

(2.206)

where, the identification with minus kinetic electronic energy comes out from the well-known virial first theorem. [25]

Certain, having the expression 2.205 for *HK* functional is not enough to assert that this has the universal properties. Firstly, one has to separate the kinetic and electronic repulsion components, as the relation 2.52 prescribes, and then to impose to the V_{ee} term the self-interaction condition,

$$V_{ee}(N \to 1) \to 0$$

(2.207)

in the same way as the 2.43 Fermi-Amaldi term requires.

The kinetic energy was already expressed in 2.206. The remaining electronic repulsion energy can be further separated out employing the second virial theorem which, for the electronic systems at equilibrium, identifies the total potential energy with the negative of the doubled kinetic one,

$$V_{ee} + C_A = -2T$$

from where the electronic repulsion energy is found to be:

$$V_{ee} = 2F_A + 2F_B + C_A$$

(2.208)

when the connection 2.206 is included.

Before to going further, worth to check the condition 2.202 on the deduced energetic components; is directly proved out that:

$$[T, V_{ee}](N \to 0 : \rho \to 0) \to 0 .$$

(2.209)

Then, the condition 2.207 have to be considered. By applying it to the expression 2.208, and counting for the components of 2.205, an equivalent chemical action functional formulation is established:

$$C_A = \frac{\ln\left(\dfrac{a+1}{a}\right) - 2\dfrac{b+1}{\sqrt{a}}\arctan\left(\dfrac{1}{\sqrt{a}}\right)}{1 + \ln\left(\dfrac{a+1}{a}\right)}$$

(2.210)

with the help of which, the electronic repulsion energy 2.208 takes the form:

$$V_{ee} = 2\frac{b+N}{\sqrt{a}}\arctan\left(\frac{N}{\sqrt{a}}\right) - \ln\left(\frac{a+N^2}{a}\right)$$

$$+ \frac{1 + \ln\left(\dfrac{a+N^2}{a}\right)}{1 + \ln\left(\dfrac{a+1}{a}\right)}\left[\ln\left(\frac{a+1}{a}\right) - 2\frac{b+1}{\sqrt{a}}\arctan\left(\frac{1}{\sqrt{a}}\right)\right].$$

(2.211)

Is now obvious that expression 2.211 well recovers the condition 2.207. However, the same relation 2.210 have to be implemented also in the kinetic 2.206 definition, giving also the negative reduced total energy functional, as:

$$-E^R = T = -\frac{b+N}{\sqrt{a}}\arctan\left(\frac{N}{\sqrt{a}}\right) + \frac{1}{2}\ln\left(\frac{a+N^2}{a}\right)$$

$$- \frac{1 + \dfrac{1}{2}\ln\left(\dfrac{a+N^2}{a}\right)}{1 + \ln\left(\dfrac{a+1}{a}\right)}\left[\ln\left(\frac{a+1}{a}\right) - 2\frac{b+1}{\sqrt{a}}\arctan\left(\frac{1}{\sqrt{a}}\right)\right].$$

(2.212)

Nevertheless, worth now to check again the condition 2.209:

$$[T, V_{ee}](N \to 0 : \rho \to 0) \to C_A(\rho \to 0) \to 0.$$

(2.213)

Last step regards the partial *HK* functional and its universal nature. Firstly, its working expression is recuperated by summing up the kinetic and the electron repulsion energies, the expressions 2.212 and 2.211, respectively, to give the functional:

$$F_{HK}^P = \frac{b+N}{\sqrt{a}} \arctan\left(\frac{N}{\sqrt{a}}\right) - \frac{1}{2}\ln\left(\frac{a+N^2}{a}\right)$$

$$\times \left[1 + \frac{2\dfrac{b+1}{\sqrt{a}}\arctan\left(\dfrac{1}{\sqrt{a}}\right) - \ln\left(\dfrac{a+1}{a}\right)}{1 + \ln\left(\dfrac{a+1}{a}\right)}\right].$$

(2.214)

Finally, the universal nature of the *HK* functional demands the non-potential dependency. Therefore, one has to switch off the potential appearance so that only the electronic density to remain in 2.214 formulation. In doing this, a high level of generality of the (V, ρ) relationship can be achieved by considering the electrostatic Poisson equation,

$$\nabla^2 V(x) = -4\pi\rho(x)$$

(2.215)

that among the initial (long-range) conditions,

$$[V(x), \nabla V(x)](x \to \infty) \to 0$$

(2.216)

provides the successive solutions:

$$-\nabla V(x) = 4\pi \int_{\infty}^{x} \rho(v)dv ,$$

$$V(x) = 4\pi \int_{\infty}^{x}\int_{w}^{\infty} \rho(v)dvdw .$$

(2.217)

Submitting the relations 2.217 successively into the 2.101, 2.187, and 2.214 ones, there is provided the only-density dependent partial *HK* functional:

$$F_{HK}^{P}[\rho] = \frac{b[\rho] + N}{\sqrt{a[\rho]}} \arctan\left(\frac{N}{\sqrt{a[\rho]}}\right) - \frac{1}{2}\ln\left(\frac{a[\rho] + N^2}{a[\rho]}\right)$$

$$\times \left[1 + \frac{2\dfrac{b[\rho] + 1}{\sqrt{a[\rho]}} \arctan\left(\dfrac{1}{\sqrt{a[\rho]}}\right) - \ln\left(\dfrac{a[\rho] + 1}{a[\rho]}\right)}{1 + \ln\left(\dfrac{a[\rho] + 1}{a[\rho]}\right)}\right],$$

$$a[\rho] = \frac{1}{4\pi} \int \frac{\nabla\rho(x)}{\int_{\infty}^{x}\rho(v)dv} dx ,$$

$$b[\rho] = \int \left[\frac{\nabla\rho(x) \int_{\infty}^{x}\int_{w}^{\infty}\rho(v)dvdw}{\int_{\infty}^{x}\rho(v)dv}\right] dx .$$

$$(2.218)$$

The partial *HK* 2.218 expression displays, indeed, a purely density functional character. However, a fully density functional analytical expression hasn't been yet formulated neither for the kinetic and repulsion terms – and consequently, nor for the entirely *HK* functional, in an universal completely manner. [16, 17, 154-162]

Nevertheless, throughout the present approach, it was succeeded in avoiding the drawback of the impossibility of expressing the kinetic and electronic repulsion terms as exact density functionals. Here, even in partial version, the obtained relation 2.218 likes to preserve the universal character associated to this functional, tracing the way in which this result was inferred.

2.3.6 The Mulliken Density Functional Electronegativity

The expression 2.186, because the way in which it was deduced, stands as the absolute electronegativity formulation, consistent with the Parr definition (see equation 2.6). However, to arrive to the Mulliken (experimentally related) electronegativity an additional specialized step is required.

Starting from Mulliken electronegativity formula, in terms of ionization potential (*IP*) and electron affinity (*EA*), equation 2.7, the relation with the Parr's electronegativity 2.6 definition can be recovered throughout the following series of identities:

$$\chi_M(N) = \frac{IP(N) + EA(N)}{2}$$

$$= \frac{[E(N-1) - E(N)] + [E(N) - E(N+1)]}{2}$$

$$= \frac{E(N-1) - E(N+1)}{2}$$

$$= -\frac{1}{2} \int_{|N-1\rangle}^{|N+1\rangle} dE_{|N\rangle}$$

$$\overset{(2.72)}{=} -\frac{1}{2} \int_{|N-1\rangle}^{|N+1\rangle} \left[\left(\frac{\partial E_{|N\rangle}}{\partial N} \right)_V dN + \int \rho_{|N\rangle}(x) dV_{|N\rangle}(x) dx \right]$$

$$= \frac{1}{2} \int_{N-1}^{N+1} \chi(N) dN - \frac{1}{2} \left[\int \rho_{|N\rangle}(x) V_{|N+1\rangle}(x) dx - \int \rho_{|N\rangle}(x) V_{|N-1\rangle}(x) dx \right].$$

(2.219)

In the virtue of the chemical action principle, 2.170 or 2.171, the two terms in the r.h.s. bracket of equation 2.219 identically vanishes, since does not optimize the pair associations of the densities with the true external potentials of the states $|N-1\rangle$, $|N\rangle$ and $|N+1\rangle$, for each such state, respectively, accordingly with the Hohenberg-Kohn theorems, see the Section 2.2.3. Therefore, the application of the chemical action principle 2.170 to 2.219 leads with the identity:

$$\chi_M(N) = \frac{1}{2} \int_{N-1}^{N+1} \chi(N) dN.$$

(2.220)

The result 2.220 was already previously founded by Komorowski adopting another way, namely the average charge electronegativity. [115, 116]

However, the present result is an exclusive density functional principles based one, i.e. on the chemical action concept and of its principle.

Further, by performing the definite integration required in the equation 2.220, with the help of the absolute electronegativity expression 2.186, one arrives to the present density functional Mulliken version of electronegativity:

$$\chi_M(N) = \frac{b+N-1}{2\sqrt{a}}\arctan\left(\frac{N-1}{\sqrt{a}}\right) - \frac{b+N+1}{2\sqrt{a}}\arctan\left(\frac{N+1}{\sqrt{a}}\right)$$

$$+ \frac{C_A - 1}{4}\ln\left[\frac{a+(N-1)^2}{a+(N+1)^2}\right].$$

$$(2.221)$$

For a given potential and for the associated electronic density, the Mulliken electronegativity functional 2.221 can be specialized both at the atomic and to the molecular levels.

To check the reliability of the Mulliken electronegativity given by relation 2.221, the bonding limit, as that of 2.189, have to be explored also in this case.

Because the Mulliken electronegativity 2.221 is a more complex than the absolute 2.186 one, the long range limit behavior $x\rightarrow\infty$ will be performed in two steps. The first one consists in the simply extraction of the limit $N\rightarrow 0$ from the genuine expression 2.221:

$$\lim_{N\to 0}\chi_M = \frac{b-1}{2\sqrt{a}}\arctan\left(-\frac{1}{\sqrt{a}}\right) - \frac{b+1}{2\sqrt{a}}\arctan\left(\frac{1}{\sqrt{a}}\right)$$

$$= -\frac{1}{2\sqrt{a}}\left[\arctan\left(-\frac{1}{\sqrt{a}}\right) + \arctan\left(\frac{1}{\sqrt{a}}\right)\right] + \frac{b}{2\sqrt{a}}\left[\arctan\left(-\frac{1}{\sqrt{a}}\right) - \arctan\left(\frac{1}{\sqrt{a}}\right)\right].$$

$$(2.222)$$

Then, on the expression 2.222 is further employed the use of the Poisson equation 2.215 within the long-range conditions 2.216. These ones allow to write the corresponding finite difference relationships:

$$\nabla V(x) \cong -4\pi\rho(x)\Delta x,$$

$$V(x) \cong 4\pi\rho(x)[\Delta x]^2.$$

$$(2.223)$$

With the help of equations 2.223, the individual terms 2.187 can be re-arranged as:

$$a \cong \sum_i^N \frac{\nabla \rho_i}{4\pi \rho_i \Delta x_i} \Delta x_i$$

$$= \frac{1}{4\pi} \sum_i^N \frac{\nabla \rho_i}{\rho_i}$$

$$\cong \frac{1}{4\pi} \sum_i^N \frac{\Delta \rho_i}{\rho_i \Delta x_i}$$

$$= \frac{1}{4\pi} \sum_i^N \frac{\Delta \rho_i}{\Delta \rho_i}$$

$$= \frac{N}{4\pi},$$

$$b \cong \int \frac{\nabla \rho}{4\pi \rho \Delta x} 4\pi \rho [\Delta x]^2 dx$$

$$= \int \nabla \rho \, \Delta x dx$$

$$\cong \int \frac{\Delta \rho}{\Delta x} \Delta x dx$$

$$= \int \Delta \rho(x) dx$$

$$= \Delta N$$

(2.224)

where, in the final line of the b term of 2.224, also the equation 2.67 was used.

The main result regards especially the approximation of $b \cong \Delta N$, as showed above. This, because by inserting this approximation back in Mulliken long-range limit 2.222, there are immediately recognized the *atoms in molecules* electronegativity formulations, so to speak, the equation 2.10 and the first order expansion of equation 2.33, respectively:

$$\chi_{\langle\rangle} = \chi - 2\eta\Delta N$$

$$\cong \chi - \gamma\chi \, \Delta N$$

(2.225)

with the actual identified components:

$$\chi \cong 0,$$

$$2\eta \cong \gamma\chi \cong 2\sqrt{\frac{\pi}{N}} \, \arctan\left(2\sqrt{\frac{\pi}{N}}\right) > 0$$

(2.226)

once the equations 2.222 and 2.225 are one-to-one related through relations 2.224.

The obvious cancellations of individual atomic χ, together with the positively atomic η, in molecules, reinforce the natural conclusion that the present Mulliken density functional formulation 2.221 supports the isolated as well as the binding atomic frameworks through the χ-equalization and η-maximum principles, see the Sections 2.2.2 and 2.2.3, respectively.

2.3.7 About the Chemical Hardness Functional Formulations

An electronic (bonded) system can be seen both like a *gas* as well as a *fluid* one.

It displays a manifestly gas nature to the frontier, so to speak, where the bond influence decreases its action, and therefore a manifestly fluid nature to the inner shell structure. The electronic gas nature at the frontier and beyond was intensively studied in the previous sections relating the long range limit:

$$[V(x), \nabla V(x)](x \to \infty) \to 0,$$

$$N \to 0: \rho \to 0: \nabla\rho \to 0: \dots$$

(2.227)

applied to the absolute electronegativity 2.186-2.189 and to its Mulliken version 2.222-2.226. This "gaseous" approach was the first natural choice, being the electronegativity concept primarily related to the frontier properties of the electronic systems and to the potency of the binding processes to be promoted.

But what about the "fluidic" approach of electronic systems? In other words, what effect will lead the inner (short range) asymptotic charge limit:

$$N \to \infty$$

(2.228)

when applies to electronegativity?

However, aiming to enlarge the treatment, the limit 2.228 worth to be applied also to an electronic system's quantity directly derived from electronegativity and which closely accounts for the inner stability, inhibits the binding, and favors the strong-correlation fluid effects. Such an electronic system's measure exists and identifies with chemical hardness 2.9. Therefore, the strong-correlation limit 2.228 will be applied, not directly to electronegativity but, to chemical hardness derived from electronegativity. The next natural step takes account of the fact that there are already considerate two type of electronegativity: the differential or the so called absolute one 2.6, with the analytical density functional expression 2.186, and the integral or the averaged one corresponding to the Mulliken electronegativity 2.7, with density functional realization 2.221. Consequently, as well the respective two related chemical hardness options arise, as displayed in the Table I.

Table I. *Electronegativity (left column) and its chemical hardness (right column) in the absolute (upper row) and the Mulliken (down row) formulations.*

χ	η
$\chi = -\int_{0}^{N} \frac{1}{S} dN - \frac{1}{S} \int s(x) V(x) dx$ $\qquad$ $\eta^{\chi} = -\frac{1}{2} \left(\frac{\partial \chi}{\partial N} \right)_{V}$	
$\chi_M = \frac{1}{2} \int_{N-1}^{N+1} \chi dN$ $\qquad\qquad$ $\eta^{\chi_M} = -\frac{1}{2} \left(\frac{\partial \chi_M}{\partial N} \right)_{V}$	

With the help of the electronegativity, as absolute and Mulliken formulas 2.186 and 2.221, respectively, the chemical hardness density functionals spring out according with the correspondences from Table I and look like:

$$\eta^{\chi} = \frac{1}{2(a+N^2)} + \frac{(a-N^2) C_A - 2Nb}{2(a+N^2)^2} ,$$

$$\eta^{\chi_M} = \frac{N^4 + 2N^2(a-1) + (a+1)^2}{4\sqrt{a}\left[a + (N-1)^2\right]\left[a + (N+1)^2\right]}\left[\arctan\left(\frac{N+1}{\sqrt{a}}\right) - \arctan\left(\frac{N-1}{\sqrt{a}}\right)\right]$$

$$+ \frac{(1 + a - N^2)C_A - 2Nb}{2\left[a + (N-1)^2\right]\left[a + (N+1)^2\right]}.$$

(2.229)

Obviously, the two chemical hardness functionals in 2.229 have quite different expressions.

Note that the strong-correlation (fluidic) limit 2.228 was not yet consider. So that, it will be further taken, respecting the N-dependencies, on both the electronegativities 2.186 and 2.221 as well as on the chemical hardnesses 2.229 expressions, with the respectively results drawn in the Table II.

Table II. *The non-local strong-correlation limits of the electronegativity (left column) and of the chemical hardness (right column) in the absolute (upper row) and Mulliken (down row) formulations.*

χ	η
$\chi^{(2)} = \left(\dfrac{1}{N} - \dfrac{\pi}{2\sqrt{a}}\right) - \dfrac{1}{N}\left(C_A + \dfrac{b}{N}\right)$	$\eta^{\chi(2)} = \dfrac{1 - C_A}{2N^2} - \dfrac{b}{N^3}$
$\chi_M^{(3)} = \left(\dfrac{1}{N} - \dfrac{\pi}{2\sqrt{a}}\right) - \dfrac{1}{N}\left(C_A + \dfrac{b}{N}\right)$	$\eta^{\chi_M(3)} = \dfrac{1 - C_A}{2N^2} - \dfrac{b}{N^3}$

Firstly, the results from the Table II clearly indicates that always there can be founded a proper strong-correlation degree of the limit 2.228 such that the absolute and the Mulliken electronegativities and the subsequent associated chemical hardnesses expressions to equalize among them.

This is an important result of the present approach proving its internal analytical consistency.

The mathematical-physical interpretation of the above results is simple and direct: as far as the charge accumulation increases in the electronic system, its average (the integral χ_M-Mulliken) picture reduces to the one-point (the derivative χ-absolute) approach, within the strong-correlation 2.228 limit.

More clear, in the strong correlation fluidic limit, the electronic system behaves, as a whole, like an unique bounded particle that precludes any further electronic acquisitions. In other words, being the inner shells filled, no other electrons are called. This last fact can be easily observed if, over the unified expressions of Table II, the limit 2.228 is further, one more time, applied, and leaves the expected results:

$$\lim_{N \to \infty} \chi^{(2)} = \lim_{N \to \infty} \chi_M^{(3)} = -\frac{\pi}{2\sqrt{a}} < 0,$$

$$\lim_{N \to \infty} \eta^{\chi^{(2)}} = \lim_{N \to \infty} \eta^{\chi_M^{(3)}} = 0.$$

(2.230)

From 2.230 follows that the strong-correlated system's electronegativity gets a negative numerical value predicting so, in no way the tendency of the next electronic acquisitions.

Moreover, the above 2.230 results correctly state that the two times strong-correlated electronegativity is accompanied by zero chemical hardness value that definitely favors the (virtual) back expansion of the strong fluidic systems to a achieve also a gaseous (reactive) character.

Therefore, the electronegativity and the chemical hardness go "hand in hand", in opposite phenomenological directions, to balance between the gaseous and fluidic characters of the bonded electronic systems.

However, one problem remains to be addressed.

As there was previous introduced, the chemical hardness can be expressed also as the inverse of the chemical softness, see equation 2.94. Thus, using the global softness expression 2.106, the related chemical hardness takes the form:

$$\eta^S = \frac{1}{2(a + N^2)}$$

(2.231)

which, again, drastically differs from those electronegativity related, see equations 2.229. Nevertheless, the expression 2.231 is the most simple one.

If one considers the limit 2.228 directly applied on 2.231,

$$\eta^{S(3)} = \frac{1}{2N^2}$$

(2.232)

the result still does not recuperate those electronegativity related, as displayed in the Table II. Only when the limit 2.228 is one more time taken over 2.232, the result finally coincides with that ones prescribed by chemical hardness values 2.230.

As a last remark, even that for electronegativities the long-range limit 2.227 was previously explored, it seems harmful to be applied also on the chemical hardness. This because, the electronegativity influence is manifestly also outside of the occupied electronic shells whereas the chemical hardness merely addresses the inner shell influence and its resistance to reactivity.

Concluding, it can be said that the softness-related 2.231 chemical hardness fits with the actual electronegativity-related ones, see Table II, only in the very strong (two times – charge correlation) applied 2.228 limit.

2.3.8 Markovian Path Integral Electronic Effective Densities

The computation of the above introduced density functionals requires the knowledge of the electronic density under the external potential influence. In this work the computation of the electronic density will be carried out within the exposed Feynman-Kleinert *PI* formalism, by using the following expressions:

$$\rho_1(x_0) = Z_1^{-1} \frac{1}{\sqrt{2\pi \, \hbar^2 \beta / m_0}} \exp[-\beta \, W_1(x_0)],$$

$$Z_1 = \frac{1}{\sqrt{2\pi \, \hbar^2 \beta / m_0}} \int_{-\infty}^{+\infty} dx_0 \, \exp[-\beta \, W_1(x_0)]$$

$$(2.233)$$

and where, the path influence was comprised within the introduced Feynman centroid x_0, see Section 2.2.6. Worth to note that the electronic density 2.233 fulfills the normalization condition:

$$\int_{-\infty}^{+\infty} \rho_1(x_0) dx_0 = 1$$

$$(2.234)$$

instead of 2.50. However, relation 2.234 does not contradict the *DFT* theorems as far as the evaluation of electronegativity and of the related quantities are envisaged. Is noted the fact that the electronegativity characterizes the whole system at the frontier limit, i.e. at the valence or the outer electronic shell. Often this frontier shell is characterized in terms of the effective electron that can be added or released from the valence shell according with the electronegativity tendency. Therefore, the condition 2.234, used to characterizes the effective valence electron behavior, is well justified.

In the general Section 2.2.5, through the discussion about the density matrix description by means of the path integrals there was shown how the memory effects can be cute off by imposing of the so-called Markovian condition,

$$\hbar\beta \to 0$$

(2.235)

with the merit that cancels the low temperature quantum fluctuations. [25] Due to the temporal nature of the quantum statistical quantity $\hbar\beta \propto \Delta t$, the limit 2.235 corresponds also to the ultra-short correlation of the involved electrons with the applied external potential. This means that, since initially the free motion of the electrons in absence of an external potential ($\Delta t = 0 \Leftrightarrow \beta = 0$) is assumed, as far as the external potential is then applied, an immediate orbit stabilization of the electronic system is reached ($\Delta t \to 0 \Leftrightarrow \beta \to 0$). In other words, the escape (unstable) paths are precluded. Finally, this limit introduces also *correlation effects* with the medium. Therefore, worth to apply the limit 2.235 also to the *PI* Feynman-Kleinert results, see equations 2.161-2.164. The smeared out potential 2.163, by changing the variable in such a way that:

$$z(x'_0) = \frac{(x'_0 - x_0)}{\sqrt{2a^2(x_0)}},$$

$$dz(x'_0) = \frac{dx'_0}{\sqrt{2a^2(x_0)}}$$

(2.236)

it can be rewritten in terms of the so called Wigner expansion, [163] of a high temperature limit 2.235, successively as:

$$V_{a^2}(x_0) = \frac{1}{\sqrt{2\pi\, a^2(x_0)}} \int_{-\infty}^{+\infty} V(x'_0) \exp\left[-\frac{(x'_0 - x_0)^2}{2a^2(x_0)}\right] dx'_0$$

$$= \frac{1}{\sqrt{2\pi\, a^2(x_0)}} \sqrt{2a^2(x_0)} \int_{-\infty}^{+\infty} V\left(x_0 + \sqrt{2a^2(x_0)}z\right) \exp\left(-z^2\right) dz$$

$$= \frac{1}{\sqrt{\pi}} \int_{-\infty}^{+\infty} \left\{ V(x_0) + \sqrt{2a^2(x_0)} z V'(x_0) + \frac{1}{2}(2a^2(x_0)) z^2 V''(x_0) + .. \right\} \exp\left(-z^2\right) dz$$

$$\cong \frac{1}{\sqrt{\pi}} \int_{-\infty}^{+\infty} \left\{ V(x_0) + \frac{1}{2}(2a^2(x_0)) z^2 V''(x_0) \right\} \exp[-z^2] dz$$

$$= V(x_0) + \frac{1}{2} a^2(x_0) V''(x_0).$$

$$(2.237)$$

Now, within the same 2.235 limit, the parameters 2.164 also become:

$$\Omega^2(x_0) \cong \frac{1}{m_0} V''(x_0),$$

$$a^2(x_0) \cong \hbar^2 \frac{\beta}{12 m_0}.$$

$$(2.238)$$

With equations 2.237 and 2.238 back into the potential 2.162 one gets:

$$W_1(x_0) \cong V(x_0) + \frac{1}{\beta} \ln\left[\frac{\sinh\left(\frac{\hbar\beta}{2} \sqrt{\frac{V''(x_0)}{m_0}} \right)}{\frac{\hbar\beta}{2} \sqrt{\frac{V''(x_0)}{m_0}}} \right]$$

$$(2.239)$$

from which, appears that the Feynman-Kleinert *PI* constrained-search algorithm in the markovian limit provides an efficient recipe to compute electronic densities using only the external potential dependence.

However, the resulting W_1 markovian potential 2.239 can be further considered into the limit 2.235. This last step agrees with the Parr and Yang approach, which have shown, [18] that the integral formulation of the Kohn-Sham (orbital) *DFT* arrives to the electronic density expression by performing the Wigner semi-classical expansion combined with the short time approximation (in β parameter). All the potential components around V can be formally interpreted as the *exchange-correlation PI potential* V_{XC}^{PI} of the medium. Even if this potential 2.239 can be expanded into higher orders, it will be here truncated at the second order expansion, [105] and yields:

$$W_1(x_0)_{\beta \to 0} \cong V(x_0) + \hbar^2 \frac{\beta}{24m_0} V''(x_0)$$

$$= V_{a^2 \to \frac{\beta}{12}}(x_0)$$

$$\equiv V(x_0) + V_{XC}^{PI}(x_0)$$

(2.240)

in which, the exchange-correlation *PI* potential of the medium,

$$V_{XC}^{PI}(x_0) = \hbar^2 \frac{\beta}{24m_0} V''(x_0)$$

(2.241)

corrects the classical external potential V.

2.3.9 Path Integral Pseudo-Potential Electronegativity Scale

To see how reliable is the previous *PI* markovian effective density picture worth to test it for the Mulliken electronegativity computation by using the actual 2.221 formulation.

There is obvious that for applying the present Mulliken electronegativity formula to the atomic systems, at least the core potential in which the valence electrons are evolved should be known.

However, to emphasis on the non-Coulombian dependency of the present approach another representation for the external potentials is here used.

The pseudopotential theory provides such information for each atomic system starting from the Li one. [164a]

It seems naturally to chose this way for the present computation, because in this case the pseudopotential is seen as the external potential that applies to the effective valence electron, in agreement with the density functional framework and with the reduced normalization condition 2.234.

For this purpose, the Stuttgart/Bonn group pseudo-potentials have been employed, [164b] into the electronegativity 2.221, through the relations 2.233 and 2.240, with the results displayed in the Figure 2.4. [53, 55, 164c].

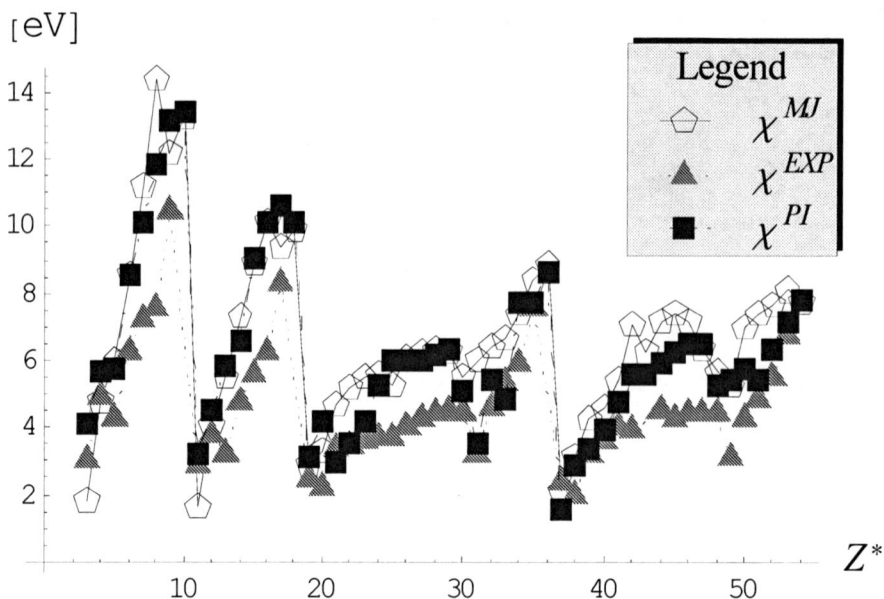

Figure 2.4. *The electronegativity (χ) versus the atomic number (Z^*) for various scales: Mulliken-Jaffe (MJ), [165a] experimental (EXP), [107] and the path-integral (PI) ones.*

As there was previously presented, the *PI* electronic density approach has its own β parametric dependency. Since the electronic density depends on the β parameter, within the *PI* calculations, this one can be fixed in such a manner that the electronic density fulfills the normalization condition 2.234 and its search must be done in the markovian limit, in atomic units: $\beta \to 0$. This proper β-rescaling procedure is able to give similar results (in absolute values) for electronegativity and the chemical action. Last assertion it nothing else but the natural generalization of the (Gordy based) previous adopted density functional formulation for the electronegativity of an electronic system under the electrostatic (Coulombian) potential of nucleus (with nuclear charge Z^*): [165b]

$$\chi(N, Z^*) = \left\langle \frac{1}{x} \right\rangle = -\int \rho(N, Z^*, x) \frac{1}{x} dx = \int \rho(N, Z^*, x) V_{COULOMB}(x) dx.$$

(2.242)

Arriving at this moment, one has to appeal the general guidelines criteria for an acceptable electronegativity scale aiming to discuss the quality of the present results. [82] According with these rules, appears that for the actual atomic electronegativity results all the major required conditions are fulfilled. The general motivations follow.

(i)   All electronegativity values measure the energy *per* electron, so that are expressed in electron volts (eV).

(ii)   Both by the general electronegativity analytical functional formulation and by the path integral implementation is assured the quantum character of the electronegativity present working scheme.

(iii)   The transitional metals display a lower range of values comparing with the main group atomic systems.

(iv)   The metalloids, namely B, Si, Ge, As, Sb, Te, Bi, in actual calculations present a narrow range of varying electronegativity.

(v)   The so called „Si rule" is valid, since all metals have the electronegativity less or equal with that of Si. However, here, a comment is necessary. The fact that Si has the lowest electronegativity value in the metalloid band seems to be not a general criteria, [82] since even in the experimental (Pearson) scale it does not apply, see both the Mulliken-Jaffe and Experimental values in the Figure 2.4 above.

(vi)   The atoms N, O, F, Ne, Ar, and Kr show the highest electronegativity values among the main groups, in each approach and also in the present path integral scale.

(vii)   The general trend for electronegativity is respected, being recorded a systematic increase left to right across rows and a general decrease down groups.

(viii) All the valence electrons has been included by appropriate effective valence electron condition 2.234. This criteria should supply the delicate problem that appears when is to be decided the number of valence electrons to be taken into account for the electronegativity computations.

However, in the present computations there was not needed to particularize the orbital type, so that only the core pseudopotential has the influence on the valence state.

This electronegativity density functional picture is a meaningful one from the physical point of view, because assumes as the basic information only the external potential, and shows also a chemical valuable character throughout the *PI* markovian Feynman-Kleinert implementation.

2.4 OUTLOOK

In the present chapter there was introduced and inferred new density functionals with a fundamental role in chemical electronic description and reactivity characterization: the chemical action, the chemical field and its period, new rate reaction functional, new functionals for electronegativity and its Mulliken *DFT* version, the electronegativity related new energetic formulations, and new functionals for the chemical hardness together with their behavior within the strong charge-correlation limit. A possible universal form for the partial Hohenberg-Kohn functional was also proposed. These functionals display a general analytical formulation in terms of the ground state density and the associated applied external potential, being suited for both the atomic and molecular implementations.

A general quantum statistical framework for the electronic density computation was presented, namely the Path Integral picture with its Feynman-Kleinert realization and its further markovian specialization. The main advantage of such enterprise is the possibility to compute the electronic density based to the only knowledge of the external potential, being this approach in close agreement with the fundamental Density Functional Theory theorems. Within such approach, the markovian limit accounts for the majority of the chemical phenomenas, in which the memory effects does not affect the long-term evolution but only the most near equilibrium state, i.e. the ground electronic states. Employing this kind of density implementation a test for atomic electronegativity scale was performed by using the present Mulliken functional particularized for the first four periods of the table of elements, being known the pseudopotential in which the effective valence electron evolves. The results follow the main accepted criteria for checking the reliability of an electronegativity scale and prove the effectiveness of the proposed *DFT+PI* combined scheme.

Being encouraged by these results, a more general reactivity characterization of the open many-electronic (molecular) systems, throughout the introduced *DFT* indices, will be in the next chapter presented.

CHAPTER 3.

THE QUANTUM-STATISTIC CHARACTERIZATION OF THE CHEMICAL REACTIVITY

Any new thing, coming from anywhere

affinities with others on the same nature…

Lucrezio T. Caro - The Nature of Things, 1112-1113

3.1 INTRODUCTION

In this chapter is given a particularization of the density functionals, introduced in the previous chapter for a general molecular model.

In order to achieve this goal, the present approach will be applied to a "pilot" system for which all the determinations will be done.

One considers a molecular electronic system in its fundamental electronic state. This will be called the state $|0\rangle$ and is assumed to be governed by the harmonic potential (in atomic units) $V_{|0\rangle} = kx^2/2$, being k the force constant. Then, this state will be "open" to interact with the environment in two different but correlated ways. Firstly, the electronic state $|0\rangle$ is vibrationally excited in the state $|g\rangle$, governed by the generalized anharmonic potential (in atomic units) $V_{|g\rangle} = kx^2/2 + gx^4/4$, being g the nuclear coupling parameter, which should be associated with an external electromagnetic g-action, changing the electronic-nuclear configuration. This (photonic) promotion will be as well particularized from weak ($g=0.1$) up to the very strong ($g=100$) vibrational (anharmonic) excitations, Figure 3.1.

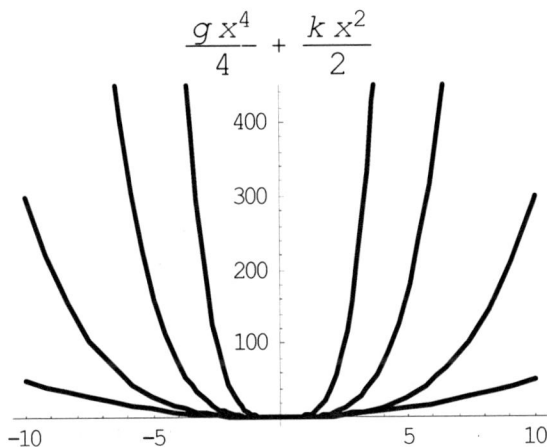

$$\frac{g\,x^4}{4} + \frac{k\,x^2}{2}$$

Figure 3.1. *Molecular potentials as harmonic (with the pair branches closer to the abscise axis) and anharmonic (with the pair branches closer to the ordinate axis) representations.*

The second kind of coupling with the external environment will be considered by means of the thermal exchange through the thermodynamic parameter β. This thermal coupling will be considered both for the fundamental electronic state $|0\rangle$ as for the (photonic) activated $|g\rangle$ ones. This way, the electronic (molecular) system is studied as an open one, since becomes subject to a chemical-thermal activation on the $|0\rangle$ ground state and to a chemical-thermo-photonic activation on the $|g\rangle$ excited vibrationally states, respectively.

The starting point lies in evaluations of the electronic densities, within the Feynman-Kleinert formalism, from the general picture into its markovian representation according with the Section 2.3.8.

Afterwards, the present reactivity indices as the density functionals, introduced in the previous chapter, will be here evaluated and represented on the (g, β) bi-variance of suited ranges, leaving with a new representation of the open molecular quantum statistical reactivity.

3.2 WORKING PATH INTEGRAL IMPLEMENTATION

In the present section all the determinations and approximations will be made in atomic units. The atomic units are defined such that: $\hbar = e = m_0 = 1/(4\pi\varepsilon_0) = 1$. This way, the atomic unit (1a.u.) is likewise a measurement unit for energy (1a.u.=4.36×10^{-18} J=27.21eV=219474.64 cm^{-1} =2625.50 kJ/mol), a unit for length – named Bohr radius (1a.u.=1a_0=0.53 $\overset{0}{A}$), a time unit (1a.u.=2.28×10^{-17} s) and also a mass unit (1a.u.=9.11×10^{-31} kg). [166]

One begins with the evaluation of the smeared out potential 2.163 for the general harmonic and anharmonic potentials:

$$V_{|0\rangle}(x) = \frac{1}{2}kx^2,$$

$$V_{|g\rangle}(x) = \frac{1}{2}kx^2 + \frac{1}{4}gx^4$$

$$(3.1)$$

leaving with the respective results:

$$V_{a^2(x)|0\rangle}(x) = \frac{k}{2}\left[x^2 + a^2(x)\right],$$

$$V_{a^2(x)|g\rangle}(x) = \frac{1}{4}\left[3g\,a^4(x) + 2kx^2 + gx^4 + 2a^2(x)(k + 3gx^2)\right].$$

$$(3.2)$$

The potentials 3.2 are to be further specialized within the markovian limit 2.238,

$$a^2(x) \cong \frac{\beta}{12}$$

and, together with the whole approximated potential 2.240 will release the quantum statistical partition functions 2.233, respectively as:

$$Z_{|0\rangle} = \frac{1}{\sqrt{2\pi\beta}} \int_{-\infty}^{+\infty} dx \exp[-\beta W_1(x)]$$

$$\cong \frac{1}{\sqrt{2\pi\beta}} \int_{-\infty}^{+\infty} dx \exp[-\beta \, V_{a^2(x)|0\rangle}(x, a^2 \rightarrow \beta/12)]$$

$$= \frac{1}{\beta\sqrt{k}} \exp\left[-k\frac{\beta^2}{24}\right],$$

$$Z_{|g\rangle} = \frac{1}{\sqrt{2\pi\beta}} \int_{-\infty}^{+\infty} dx \exp[-\beta W_1(x)]$$

$$\cong \frac{1}{\sqrt{2\pi\beta}} \int_{-\infty}^{+\infty} dx \exp[-\beta \, V_{a^2(x)|g\rangle}(x, a^2 \rightarrow \beta/12)]$$

$$= \frac{1}{4} \sqrt{\frac{4k+g\beta}{\pi \, g\beta}} K_{\frac{1}{4}}\left[\frac{\beta(4k+g\beta)^2}{128g}\right] \exp\left[\frac{\beta(48k^2+8gk\beta+g^2\beta^4)}{384g}\right]$$

(3.3)

being $K[]$ the modified Bessel function of the second rank.

With the help of the partition functions 3.3, in the markovian limit, the associated electronic densities 2.233 will be taken as (see also the Section 3.4):

$$\rho_{|0\rangle}(x) \cong Z_{|0\rangle}^{-1} \exp\left[-\frac{\beta \, k}{2} x^2\right],$$

$$\rho_{|g\rangle}(x) \cong Z_{|g\rangle}^{-1} \exp\left[-\beta\left(k\frac{x^2}{2} + g\frac{x^4}{4}\right)\right].$$

(3.4)

Then, the expressions for the total number of electrons, see equation 2.50, are given as:

$$N_{|0\rangle}(x) = \int_{-\infty}^{+\infty} \rho_{|0\rangle}(x)dx$$

$$= \sqrt{2\pi\beta}\,\exp\left[\frac{k\beta^2}{24}\right],$$

$$N_{|g\rangle}(x) = \int_{-\infty}^{+\infty} \rho_{|g\rangle}(x)dx$$

$$= 2\sqrt{\frac{2\pi\beta\,k}{4k+g\beta}}\,\frac{K_{\frac{1}{4}}\left[\dfrac{\beta\,k^2}{8g}\right]}{K_{\frac{1}{4}}\left[\dfrac{\beta\,(4k+g\beta)^2}{128g}\right]}\exp\left[-\frac{\beta^2(8k+g\beta^3)}{384}\right].$$

(3.5)

The electronic densities 3.4 together with the original external potentials 3.1 immediately provide the chemical actions expressions, according with the basic definition 2.53, to be:

$$C_{A|0\rangle}(x) = \int_{-\infty}^{+\infty} \rho_{|0\rangle}(x)V_{|0\rangle}(x)dx$$

$$= \sqrt{\frac{\pi}{2\beta}}\,\exp\left[\frac{k\beta^2}{24}\right],$$

$$C_{A|g\rangle}(x) = \int_{-\infty}^{+\infty} \rho_{|g\rangle}(x)V_{|g\rangle}(x)dx$$

$$= 2\sqrt{\frac{2\pi}{4k+g\beta}}\,\frac{\exp\left[-\dfrac{\beta\left(48k^2+8gk\beta+g^2\beta^4\right)}{384g}\right]}{(g\beta)^{5/4}\,K_{\frac{1}{4}}\left[\dfrac{\beta\,(4k+g\beta)^2}{128g}\right]}$$

$$\times \left\{ \sqrt{g\beta}\ \Gamma\left[\frac{5}{4}\right]\left[g\ _1F_1\left(\frac{5}{4},\frac{1}{2},\frac{k^2\beta}{4g}\right) - k^2\beta\ _1F_1\left(\frac{5}{4},\frac{3}{2},\frac{k^2\beta}{4g}\right) \right] \right.$$

$$\left. + gk\beta\ \left[\Gamma\left[\frac{3}{4}\right]_1F_1\left(\frac{3}{4},\frac{1}{2},\frac{k^2\beta}{4g}\right) - \Gamma\left[\frac{7}{4}\right]_1F_1\left(\frac{7}{4},\frac{3}{2},\frac{k^2\beta}{4g}\right) \right] \right\}$$

(3.6)

being here $\Gamma[\ ]$ and $_1F_1()$ the Gamma Euler and the hyper-geometric functions, respectively.

In order to evaluate the electronegativity, the hardness and the energetic functionals, the sensitivity components 2.187 and the non-local term 2.101 have to be firstly estimated. The analytical expression for the function 2.101 will, respectively, take the forms:

$$L_{|0\rangle}(x) = \sqrt{k}\beta^2 \exp\left[-\frac{k\beta\left(12x^2 - \beta\right)}{24} \right],$$

$$L_{|g\rangle}(x) = 4\sqrt{\frac{\pi\ g\ \beta^3}{4k + g\beta}}\ \frac{\exp\left\{ -\dfrac{\beta\left[48k^2 + 8gk\left(24x^2 + \beta\right) + g^2\left(96x^4 + \beta^4\right)\right]}{384g} \right\}}{K_{\frac{1}{4}}\left[\dfrac{\beta\left(4k + g\beta\right)^2}{128g} \right]}$$

(3.7)

with the help of which, the terms of the type 2.187 become:

$$a_{|0\rangle}(x) = \int_{-\infty}^{+\infty} \frac{\nabla\rho_{|0\rangle}(x)}{\left[-\nabla V_{|0\rangle}(x)\right]} dx$$

$$= \sqrt{2\pi\beta^3}\ \exp\left[\frac{k\beta^2}{24} \right],$$

$$a_{|g\rangle}(x) = \int_{-\infty}^{+\infty} \frac{\nabla\rho_{|g\rangle}(x)}{\left[-\nabla V_{|g\rangle}(x)\right]} dx$$

$$= 2\sqrt{\frac{2\pi \, k\beta^3}{4k + g\beta}} \frac{K_{\frac{1}{4}}\left[\dfrac{\beta \, k^2}{8g}\right]}{K_{\frac{1}{4}}\left[\dfrac{\beta \, (4k + g\beta)^2}{128g}\right]} \exp\left[-\frac{\beta^2 \left(8k + g\beta^3\right)}{384}\right],$$

$$b_{|0\rangle}(x) = \int_{-\infty}^{+\infty} \frac{\nabla\rho_{|0\rangle}(x)}{[-\nabla V_{|0\rangle}(x)]} V_{|0\rangle}(x)dx$$

$$= \sqrt{\frac{\pi\beta}{2}} \exp\left[\frac{k\beta^2}{24}\right],$$

$$b_{|g\rangle}(x) = \int_{-\infty}^{+\infty} \frac{\nabla\rho_{|g\rangle}(x)}{[-\nabla V_{|g\rangle}(x)]} V_{|g\rangle}(x)dx$$

$$= 2\sqrt{\frac{2\pi}{4k + g\beta}} \frac{\exp\left[-\dfrac{\beta \left(48k^2 + 8gk\beta + g^2\beta^4\right)}{384g}\right]}{g^{5/4}\beta^{1/4} K_{\frac{1}{4}}\left[\dfrac{\beta \, (4k + g\beta)^2}{128g}\right]}$$

$$\times \left\{\sqrt{g\beta}\ \Gamma\left[\frac{5}{4}\right]\left[g\ _1F_1\left(\frac{5}{4},\frac{1}{2},\frac{k^2\beta}{4g}\right) - k^2\beta\ _1F_1\left(\frac{5}{4},\frac{3}{2},\frac{k^2\beta}{4g}\right)\right]\right.$$

$$\left. + gk\beta\left[\Gamma\left[\frac{3}{4}\right]_1F_1\left(\frac{3}{4},\frac{1}{2},\frac{k^2\beta}{4g}\right) - \Gamma\left[\frac{7}{4}\right]_1F_1\left(\frac{7}{4},\frac{3}{2},\frac{k^2\beta}{4g}\right)\right]\right\}.$$

(3.8)

These are the basic relations, from which will be proceeded in calculating of the density functionals introduced in the Chapter 2 for the electronic states $|0\rangle$ and $|g\rangle$, as open systems, within the markovian internal (g) and environmental (β) couplings limits, for the pilot specialization $k=1$ of the force constant.

3.3 MARKOVIAN EVALUATION OF THE CHEMICAL DESCRIPTORS

3.3.1 Densities and Electronic Numbers

The natural starting point regards the electronic density and the associated analyzed number of electrons, with the representations depicted in the Figures 3.2 and 3.3, respectively.

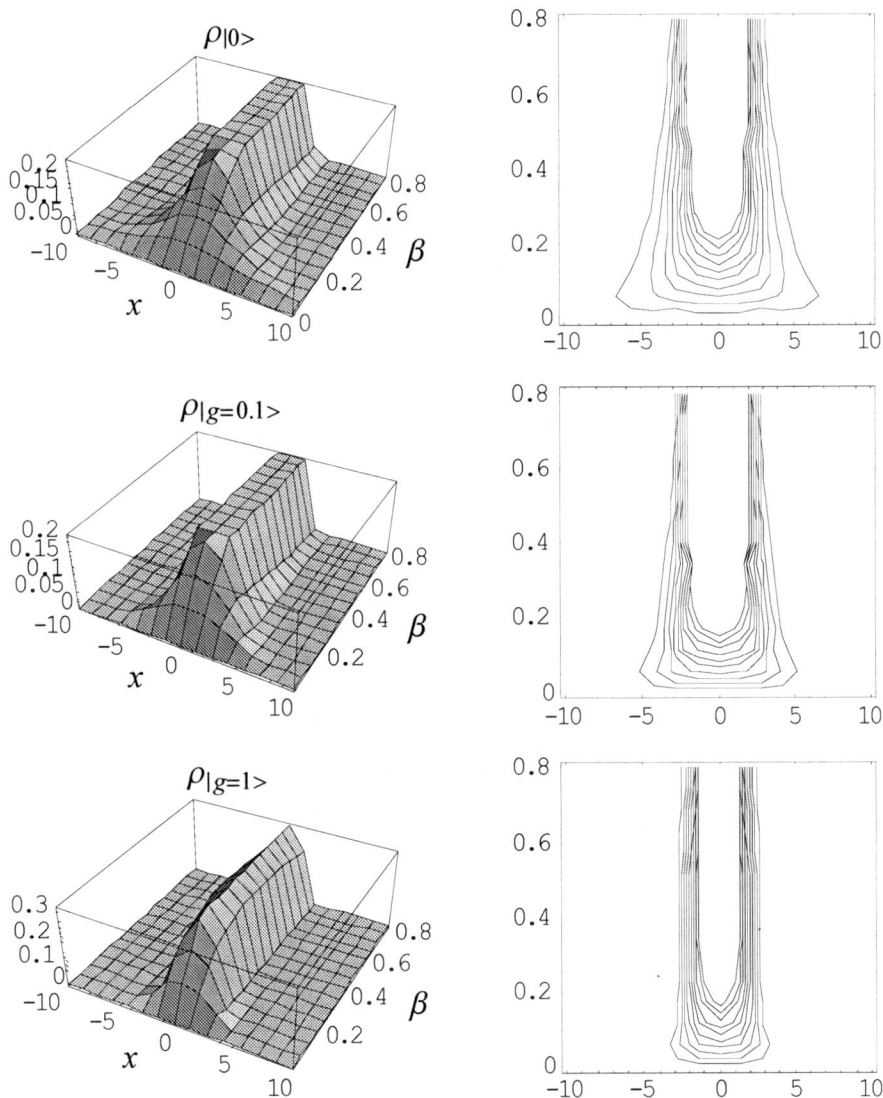

Figure 3.2. *The uni-electronic densities associated to the electronic states $|0\rangle$ and $|g\rangle$ (left) as well as their equi-contour projections (right), respectively.*

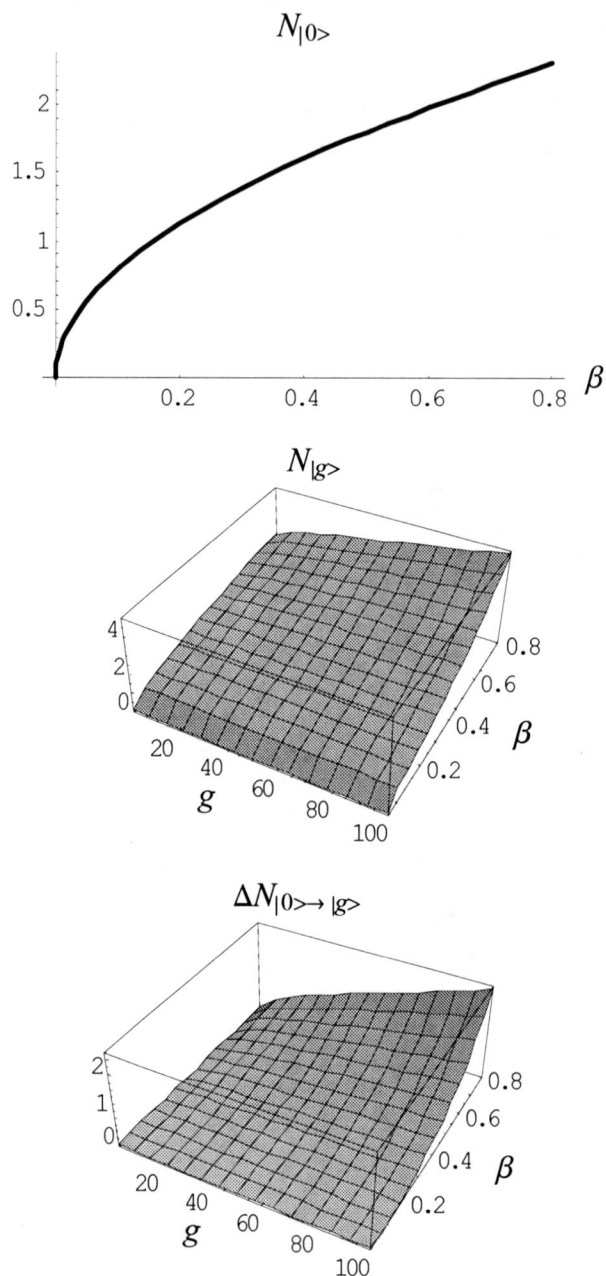

Figure 3.3. *The numbers of electrons associated to the open molecular electronic* $|0\rangle$ *and* $|g\rangle$ *states (the top and middle representations) together with their difference (in the bottom), respectively.*

By the inspection of the Figure 3.2 there is remarked that as the vibrational excitations take place the electronic density narrows around the origin of the molecular space, which shall be associated with the mass center (MC) of the nuclear system, and rises slightly in its topmost value.

Regarding the numbers of electrons in the vibrationally states, within the internal coupling ranging from very small to the very strong one: $g \in [0,100]$, from the representations displayed in the Figure 3.3 is turning out that the difference between the numbers of electrons in the electronic states $|0\rangle$ and $|g\rangle$ varies between 0 and 2.

Moreover, this maximum of two electrons, as the difference, is achieved when the environment thermal coupling (β) increases, in accordance with the quantum orbital spin-restriction, i.e. Pauli principle.

However, what interests from the chemical point of view regards the individuation of the proper couplings that produce the difference in the numbers of electrons to becomes unity. This fact makes out the right framework in treating the fundamental uni-electronic addition/promotion exchanges of the envisaged electronic systems. This observation will be used in a further paragraph for an extension of the universal Parr electronegativity atomic value, see equation 2.47, to a general definition of the Parr electronegativity for the open (molecular) electronic systems.

3.3.2 The Chemical Action Related Indices

Passing to the representation of the density functionals, the chemical action, according to the expressions 3.6, is shown in the Figure 3. 4 together with the corresponding draws of the rate reaction functional 2.191.

This way, for a chemical action, one remarks how the system in conditions of high temperature's couplings rises its energetic "response" for the molecular electrons, but does not create a good selection of certain stationary chemical actions, i.e. the chemical action principle 2.170 fulfilled. Instead, a good selection of certain constant chemical action can be reached out on the intermediary temperature's couplings, no matter at which degree of internal coupling range it appears. This fact is recorded by the Figure 3.4-left, up and down, where the chemical action displays a land with a constant tendency. Is worthy that for the environment couplings on which the chemical actions are stabilized the corresponding reaction rates are vanishing, Figure 3.4-right: up and down. As the β coupling starts to increase, the rate reaction response is very sensitive and it can pass also through critical negative rate recordings, so indicating that the electronic flux is driven from the environment to the system.

Such critical behavior states as the direct consequence of the slope changes in the chemical action shapes: compare the trends of the left-right representations in the bottom of Figure 3.4.

Consequently, the chemical action creates a potential selection of the reactivity for an open anharmonic (molecular) electronic system on its bi-variance (internal-environmental) coupling behavior.

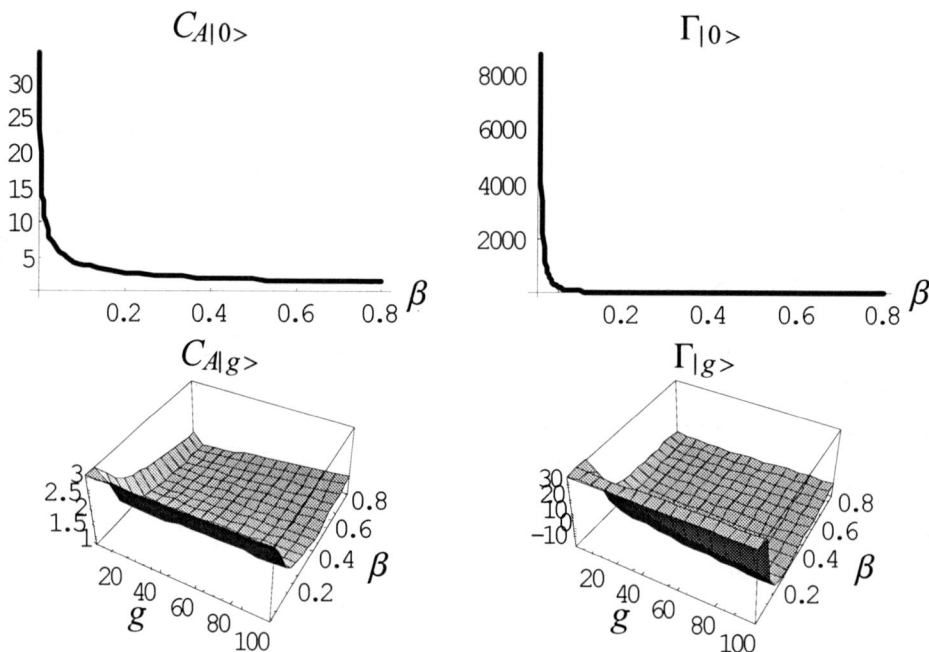

Figure 3.4. *The chemical actions (left) and their associated rate reaction responses (right) for the electronic $|0\rangle$ and $|g\rangle$ states, respectively.*

With the help of chemical action expressions 3.6, also the correspondent chemical fields and   periods, see relations 2.183 and 2.184, are   next represented and analyzed through the Figure 3.5, respectively.

Firstly, there is remarked the finite spatial domain for the chemical field representations, a behavior that certainly indicates the bond frontiers for a specific coupling conditions.

Naturally, by recording the constancy of the bonding spatial regions, the vertical amplitude of the chemical fields decreases as the vibrational coupling increases, Figure 3.5: left side movies, from top to bottom.

Since the physical dimension of the chemical field corresponds to a throb, its behavior worth to be interpreted together with those associated by the correspondent chemical periods.

Taking into account the used temporal scale, the chemical field's period is measured into femto-seconds, a range that turns to be the right temporal scale in which the chemical reactions takes places. [14]

Obviously, also the inverse phenomenology applies for the chemical periods: they parallel the vibrational excitations making those states more suitable for the laser controlled chemical reaction experiments. [33-38]

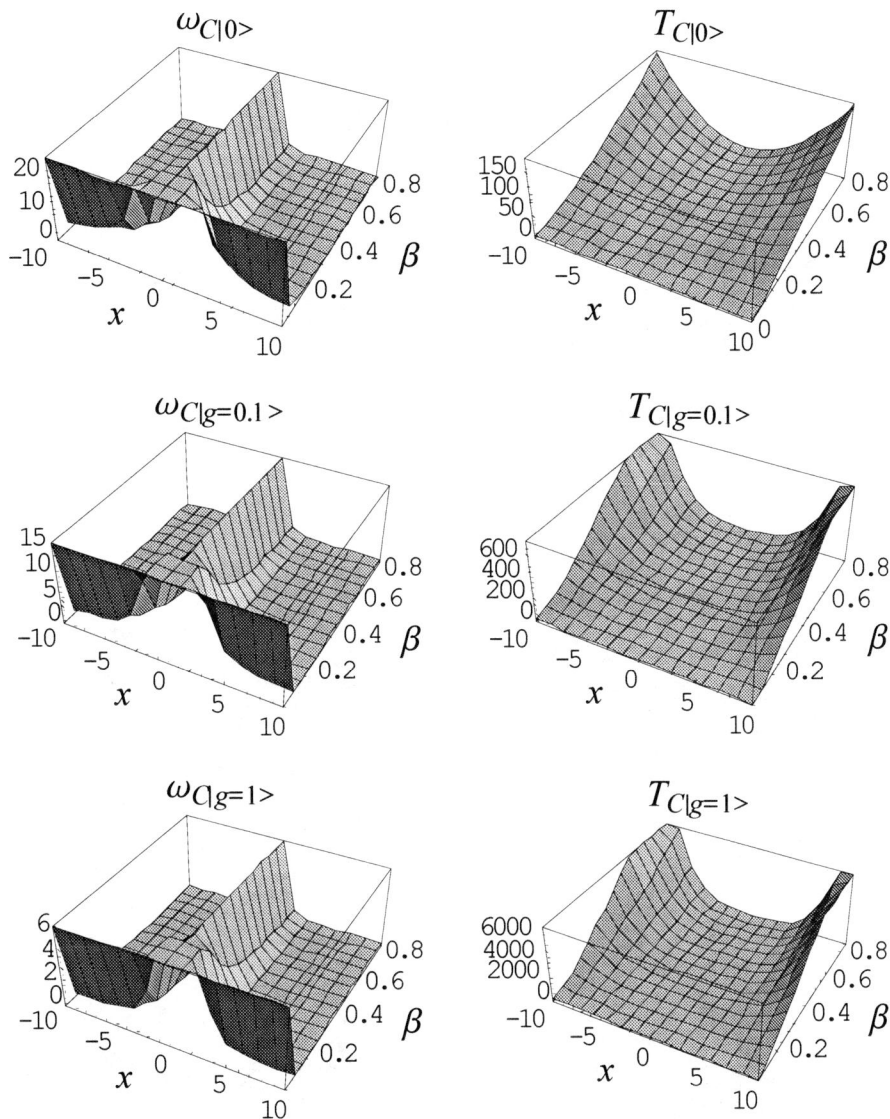

Figure 3.5. *The chemical fields (left) and their periods (right) for a molecular electronic system in the vibrational $|0\rangle$ and $|g\rangle$ states, with the internal coupling fixed to the very weak (top), weak (middle), and intermediary (bottom) excitations, respectively.*

3.3.3 The Sensitivity Indices

The discussion of the sensitivity indices starts with the representations of the local softnesses and the Fukui functions, see formulas 2.106 and 2.93, respectively depicted in the Figure 3.6.

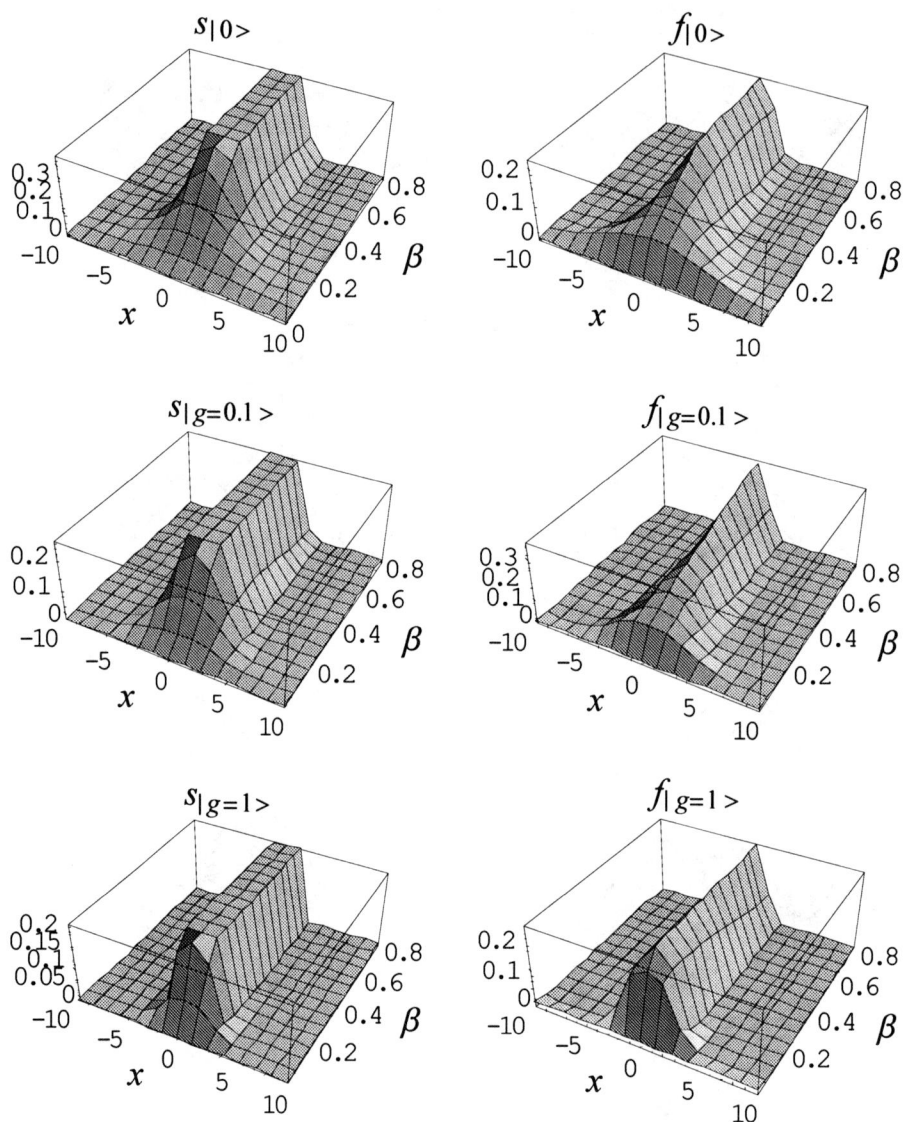

Figure 3.6. *The local chemical softness (left) and the Fukui functions (right) for a molecular electronic system in the vibrational* $|0\rangle$ *and* $|g\rangle$ *states, with the internal coupling fixed to the very weak (top), weak (middle), and intermediary (bottom) excitations, respectively.*

There are clearly remarked the previous noted spatial restricted chemical field response regions, since the local softness follows almost the same local trend as the electronic density representations does, compare left sides in Figures 3.6 and 3.2, respectively. The same phenomenology applies for the Fukui function shapes in the right sides of Figure 3.6. Worth also to emphasis on the Fukui function *quasi* regular shape response on the whole markovian environment coupling, once the electronic envisaged state is excited within an intermediary vibrationally coupling, Figure 3.6: bottom-right. Such a result accords with the previous chemical field prescription that predicts a better control of the reactivity (or the *quasi* sudden Fukui function response) when the concerned electronic states are firstly excited. This phenomenology corresponds with the consecrated pump (first excitation) – probe (final excitation) double exciting laser control of the chemical reactions, [38] and fit also with the recent "catastrophe" characterization of the chemical bonds and reactivity. [167]

Next on, the global softness is discussed in relation with its associated chemical hardness index, see relations 2.106 and 2.94, based on their representations depicted in the Figure 3.7.

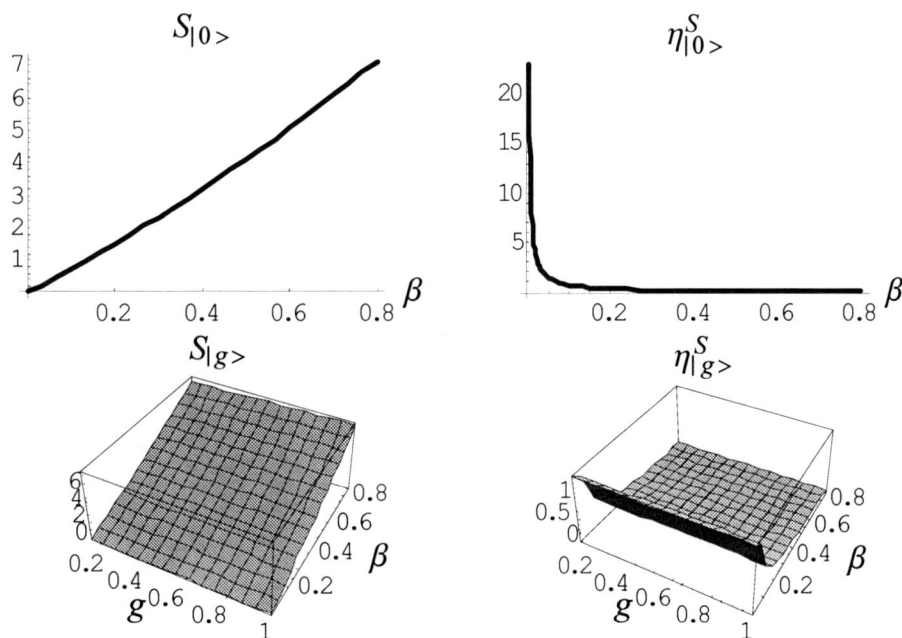

Figure 3.7. *The global chemical softness (left) and its associated chemical hardness (right) for a molecular electronic system in the vibrational* $|0\rangle$ *and* $|g\rangle$ *states, respectively.*

In the Figure 3.7 is evidenced the favorable tendency for the reactivity of the electronic states activated within intermediary temperature couplings, both as the increasing in the global softness as well as the decreasing in its corresponding chemical hardness, as inverse quantum observables. More, the all range-all couplings positively values for chemical hardness reinforces the parabolic total energy N-dependence shape, see the Figures 2.1 and 2.3 ones.

The difference between the vertical values for the softness trends is almost irrelevant among the *non-* and the respective vibrationally excited states, compare the vertical scales on the upper and down left sides of Figure 3.7.

However, the difference is noticeable for the reactivity hardness inhibition that decrease up to twenty times once the concerned electronic state is only intermediary excited ($|0\rangle \rightarrow |0.1 \div 1\rangle$) in a superior vibrational state, compare the vertical scales on upper and down left sides of Figure 3.7.

This means that, up to now, the most sensitive reactivity index of the vibrational coupling regards the chemical hardness. Let's analyze next, its reactive shape in relation with electronegativity, throughout the relation 2.9.

When the absolute electronegativity 2.186 is taken as the reference reactive index the respective chemical hardness from 2.229 will behave on the bi-variance coupling interaction as depicted in the right side of Figure 3.8.

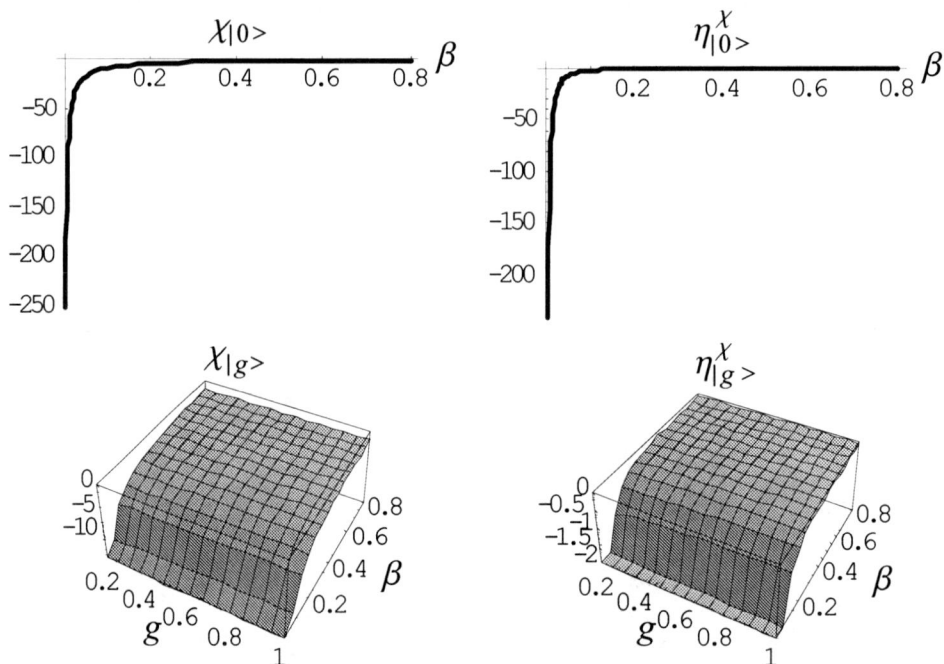

Figure 3.8. *The absolute electronegativity (left) and the associated chemical hardness (right) for a molecular electronic system in the vibrational $|0\rangle$ and $|g\rangle$ states, respectively.*

The first obvious remark relies on the same sign and trends of the represented absolute electronegativity and of related chemical hardness in Figure 3.8. The noticeable contraction of the vertical scale when the electronic state is turned into a vibrational coupled one is as well evident. Last remark fits with the correct reactivity feature on excited states, already predicted through previous analyzed reactivity descriptors. But the fact that both electronegativity and chemical hardness, even with the same tendency respecting various couplings, behaves with the same sign is not convincible due to the expected phenomenological inverse relation between these two reactivity indices.

The situation becomes more reasonable when the chemical hardness is expressed out from the Mulliken electronegativity formulation 2.221, with its analytical realization given by the respective formula from 2.229, and depicted within the concerned coupling representations in the Figure 3.9.

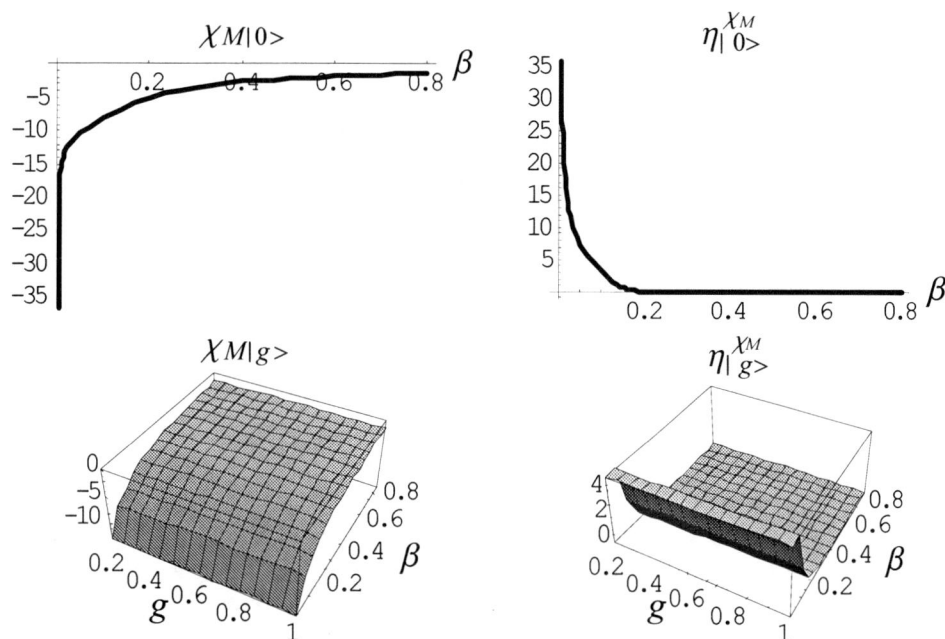

Figure 3.9. *The Mulliken electronegativity (left) and the associated chemical hardness (right) for a molecular electronic system in the vibrational $|0\rangle$ and $|g\rangle$ states, respectively.*

Again, there are maintained all the previous good features regarding the vertical scale contraction once is passing from very (very) weak to the intermediary vibrationally excited states, a fact evidenced also in Figure 3.8, and recuperated also here for both the Mulliken electronegativity and its chemical hardness representation, in the Figure 3.9. Instead, now, their trends display the correct opposite signs, having the positive chemical hardness, in right sides of Figure 3.9, the same behavior and sign likewise the softness related one, in the right sides of Figure 3.7, being this time only vertically corrected.

Lowering the temperature (or increasing the β) coupling there are recorded the constancy, and practically the zero value, approach of the electronegativity and chemical hardness trends, respectively. This effect has quantum-statistical roots, enlighten by the *quasi* free electronic fluxes at low temperature. However, for all the temperature couplings the minimum in electronegativity correlates with the maximum in chemical hardness behaviors, as the Maximum Hardness Principle (see 2.70 and 2.71) demands.

3.3.4 The Reduced Total Energy and the Partial Hohenberg-Kohn Functional

In first, the kinetic and the electron-electron repulsion energies are represented, employing the terms 3.8 and expression 3.5 into the formulation 2.212 and 2.211, respectively, with their drawings in the Figure 3.10.

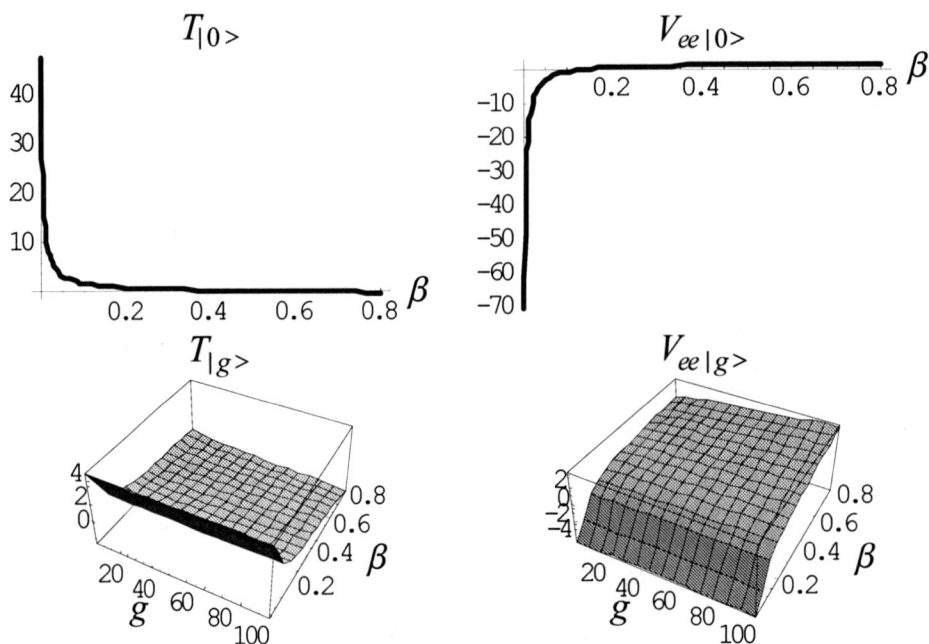

Figure 3.10. *The kinetic energy (left) and the electron-electron interaction energy (right) for a molecular electronic system in the vibrational $|0\rangle$ and $|g\rangle$ states, respectively.*

By the analysis of the displayed trends of the energetic terms depicted in the Figure 3.10 there is correctly remarked the always positively kinetic energy character that decreases with the temperature coupling relaxation or (increase in β coupling), whereas the electron-electron interaction decreases in the negative energetic range, and, at some specific temperature (coupling) changes the sign. How can be this feature explained?

There was previous mentioned the spin-coupling effect, remarked in Figure 3.3, according with which the number of electrons starts to increase on a stable with the increase in both the environment and vibrationally, β-g, couplings, respectively.

As the number of electrons approaches values over the unity appears the situations when the electronic quantum spin counts. The subsequent applied couplings cannot in no way increase this number for the same-spin electrons, being also the cases with more than two opposite-spin electrons in the same quantum state forbidden by the violation of the Pauli principle.

Therefore, at very (high) vibrational couplings, the change of sign in electron-electron repulsion arises as a natural quantum consequence, noted in the Figure 3.10, and denotes the appearance of the Fermi holes. [167]

Then, with the help of the kinetic and of the electron-electron interaction energies, the reduced total energy 2.212 and the partial Hohenberg-Kohn 2.214 can be also represented, with respectively draws displayed in the Figure 3.11.

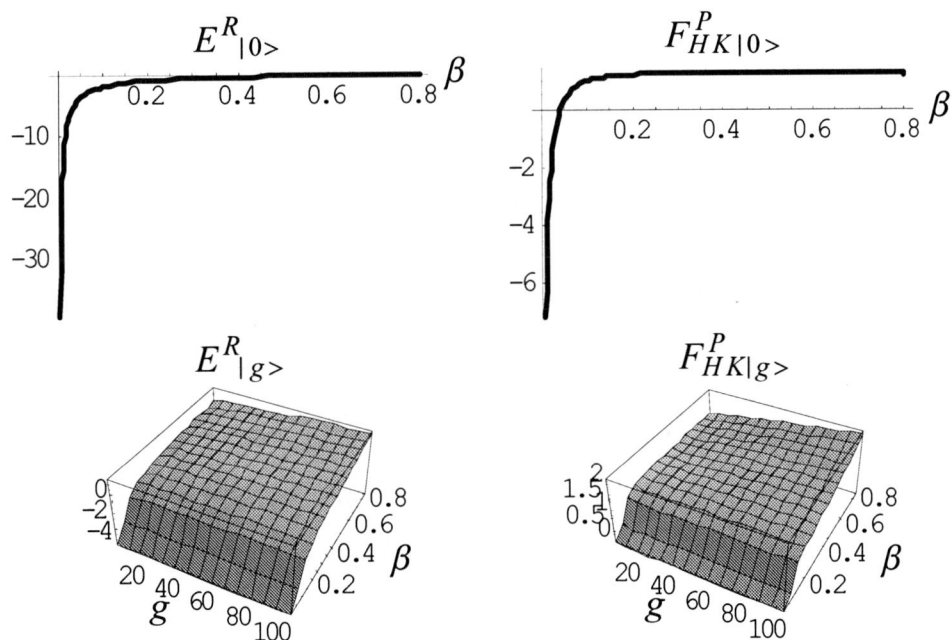

Figure 3.11. *The reduced total energy (left) and partial Hohenberg-Kohn energy (right) for a molecular electronic system in the vibrational* $|0\rangle$ *and* $|g\rangle$ *states, respectively.*

Regarding the reduced total energy reactive shapes, depicted in left sides of Figure 3.11, there are correctly obtained the negative values on their hyper-surfaces that approach a constant (quantum stabilized) level within the increasing β – coupling environment influence.

However, the partial Hohenberg-Kohn energy parallels, in the absence of the vibrational coupling, the same two sign regions as for the electron-electron interaction one, previously discussed in the right side of Figure 3.10.

More, as the vibrational coupling activation applies, it provides only the positive Hohenberg-Kohn energy values, as is recorded from the right-down representation in Figure 3.11.

Looking to the vertical values of kinetic and electron-electron interaction terms in the bottom of Figure 3.10 one realizes that they are almost equal and inversed one to other as the $1/\beta$ – coupling increases, becoming dominated by the positive values of the electron-electron interaction terms in the low temperature $(1/\beta)$ – coupling. The former behavior corresponds to the thermal activation, whereas the last one is the quantum Fermi hole response against the coupling as prescribed by the Pauli exclusion principle.

Nevertheless, for the total energy computation the chemical action term has to be considered with its 2.210 formulation, and is remarkable that this form will cancel all forbidden (the Fermi-hole) contributions in the Hohenberg-Kohn positive energetic scale when together are considerate to give the reduced total energy values, recuperating so the correct negative bonding energetic range.

3.3.5 The Parr Electronegativity

Finally, worth to introduce and evaluate the electronegativity difference between the vibrationally activated sates, $\chi_{|g\rangle}$, and that of the fundamental vibrational electronic one, $\chi_{|0\rangle}$, providing is such the appropriate information about the reliability of a given system, or the potential model, to be ascribed to a chemical process.

Therefore, a suitable driving electronegativity measure is introduced, namely the so called Parr electronegativity:

$$\chi_P \equiv \chi_{|g\rangle} - \chi_{|0\rangle}$$

$$(3.9)$$

a new reactivity index to be evaluated and represented within the ranges of vibrational and environmental previous analyzed couplings.

However, because two electronegativity formulations are available, the absolute 2.186 and the Mulliken 2.221 ones, the Parr electronegativity index 3.9 will be for both these cases discussed, throughout the drawings of Figure 3.12.

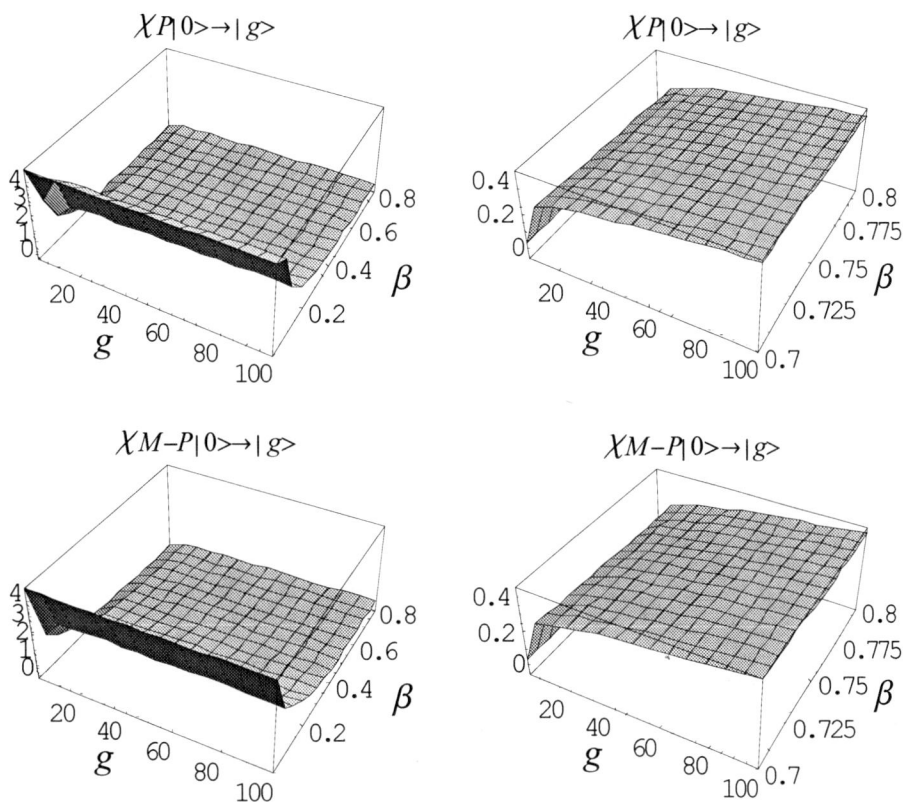

Figure 3.12. *The Parr electronegativities from the absolute (up) and from the Mulliken (down) formulations for the whole markovian (left) and for the intermediate (right) environmental coupling ranges of a molecular electronic system between the vibrational $|0\rangle$ and $|g\rangle$ states, respectively.*

From the Figure 3.12 is clear that no significant difference, neither in shape nor in vertical values, is recorded for the Parr electronegativity based on the absolute or on the Mulliken electronegativity approaches, respectively.

Such "universal" behavior makes from the definition 3.9 a valid one and consecrate it as a new reactivity index that gives the important indication for searching of the suitable shapes for which the atomic universal Parr electronegativity result, 2.47, is recovered at the molecular level.

To achieve this purpose the representation of Parr electronegativity 3.9 was redone for an intermediary environmental range, $\beta \in (0.7, 0.8)$, the range in which Parr electronegativity records the smallest vertical interval variation, see right sides in Figure 3.12. Moreover, this range of computation corresponds with the range in which the difference in numbers of electrons associated with the involved vibrationally coupled states varies around unity, see the bottom representation of Figure 3.3.

By considering these arguments, it follows that for the parametrical ranging couplings individuated as $g=10\div20$ and $\beta=0.7\div0.8$, respectively, the variation of the number of electrons is around $\Delta N = 1$ in the Figure 3.3. Mapping these ranges in the Figure 3.12 there are obtained vertical values (in a. u.) which, transposed in eV, correspond of the value 5.4 eV, just around the value established by Parr as 5.1eV, see the result 2.47, for the so called universal atomic (or orbital) electronegativity.

This result justifies the notation, interpretation and the phenomenological extension assigned to the introduced Parr electronegativity, stimulating it so far to become an operative concept for analyzing the coupled electronic systems, according to the definition 3.9, anytime when the variation of the electronic number corresponds to unity, $\Delta N=1$.

Further on, with the help of the Parr electronegativity, in its generalized 3.9 formulation, the geometrical and the arithmetical average approaches of atoms in molecules can be consider. This procedure is meaningful for the binding processes in which the vibrational and the thermal couplings give significant contributions to the molecular bonding. Therefore, here, will be mimic the bonding situation between two electronic systems: one in the fundamental vibrational $|0\rangle$ state and another in a vibrational excited (for instance, by a pump laser) $|g\rangle$ state. The first mobile object in this approach states the difference $\Delta N \equiv N_{|g\rangle} - N_{|0\rangle}$, already represented in the bottom of Figure 3.3. In the case of the geometrical average approach of the electronegativity, see equation 2.32, the bond electronegativity will be given like:

$$\chi^{\sqrt{}} = \sqrt{\chi_{|0\rangle}\chi_{|g\rangle}}$$

(3.10)

and where, the involved parameter in 2.34 is firstly determined as:

$$\gamma = -\frac{1}{2\Delta N}\ln\left[\frac{\chi_{|0\rangle}}{\chi_{|g\rangle}}\right].$$

(3.11)

With the help of the parameter 3.11, the released energy 2.35 is successively rewritten according with the concerned involved states and their Parr electronegativity, see definition 3.9, as:

$$\Delta E^{\sqrt{}} = \Delta N \left\{ \chi_{|0\rangle} - \chi_{|g\rangle} - \frac{\chi_{|0\rangle} + \chi_{|g\rangle}}{4} \ln \left[\frac{\chi_{|0\rangle}}{\chi_{|g\rangle}} \right] \right\}$$

$$= -\Delta N \left\{ \chi_P - \left(0.5 \chi_{|0\rangle} + \frac{\chi_P}{4} \right) \ln \left[1 + \frac{\chi_P}{\chi_{|0\rangle}} \right] \right\}.$$

(3.12)

The (g, β) bi-variance representation of the geometrical average approach of the bond electronegativity and of its released energy, are represented in the Figure 3.13, respectively.

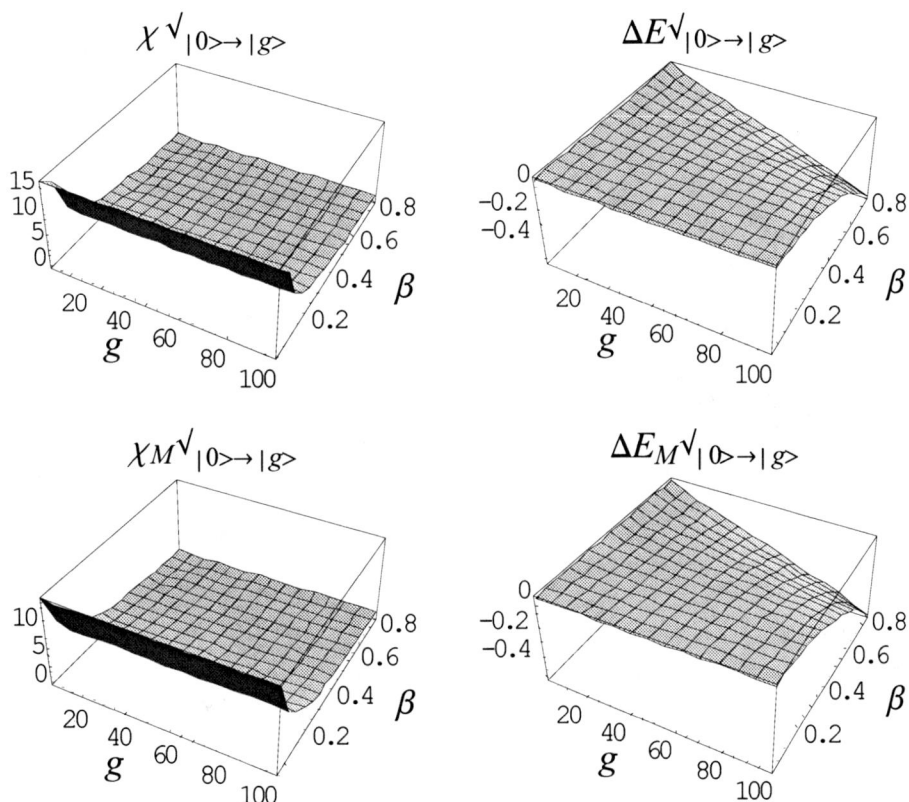

Figure 3.13. *The geometrical average approach of the bond electronegativities (left) and of the associated released energies (right), within the absolute (up) and the Mulliken (down) pictures, since the binding process is considerate between two electronic systems, being one in the vibrational $|0\rangle$ state and the other in the excited $|g\rangle$ one, respectively.*

The Figure 3.13 highlights correctly the fact that as the bonding electronegativity decreases more energy have to be released, which in turn, corresponds with the increase in the number of the electrons to be exchanged between the vibrationally $|0\rangle$ and $|g\rangle$ states, as also the bottom representation in the Figure 3.3 prescribes.

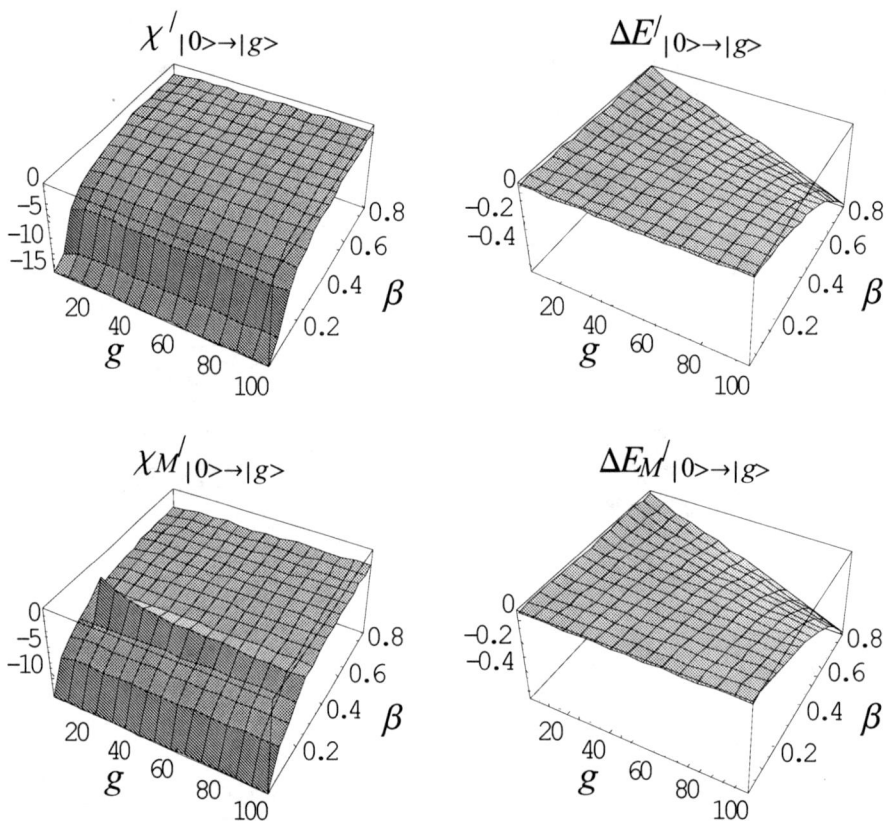

Figure 3.14. *The arithmetical average approach of the bond electronegativities (left) and of the associated released energies (right), within the absolute (up) and the Mulliken (down) pictures, since the binding process is considerate between two electronic systems, being one in the vibrational $|0\rangle$ state and the other in the excited $|g\rangle$ one, respectively.*

Passing to the arithmetic equalization approach for the binding electronegativities, based on the working relations 2.24 and 2.23, the bond electronegativity and its released energy are here evaluated, respectively, like:

$$\chi' = \frac{S_{|0\rangle}\chi_{|0\rangle} + S_{|g\rangle}\chi_{|g\rangle}}{S_{|g\rangle} + S_{|0\rangle}},$$

$$\Delta E' = \frac{\chi_{|0\rangle} - \chi_{|g\rangle}}{2}\Delta N = -\frac{\chi_P}{2}\Delta N$$

(3.13)

being their associated coupling hyper-surfaces depicted in Figure. 3.14.

To obtain 3.13 worth noting that, in order to don't more complicate the practical particularizations of the working relations, the original chemical hardnesses in expression 2.24 was replaced by its associated global softness, using the unique connection 2.94, $\eta = 1/(2S)$, instead of using the double related absolute and Mulliken electronegativity formulations through the relations 2.229.

The analysis of the Figure 3.14 reveals many meaningfully characters. Regarding the bonding electronegativity there is recorded the same general shape as the individual binding electronegativities do, see the left sides of Figure 3.9, for instance. The vertical range values of the bond electronegativity are correctly reduced respecting the simple arithmetic average of the vertical scales of the binding electronegativities from the left sides of the Figure 3.9, underlining therefore on the equilibrium stabilization. Also, the stabilization to a *quasi* constant electronegativity of the bond fits, in the Figure 3.14, with the equalization principle of electronegativity binding components, see the Sanderson and the frontier electron principles in Section 2.2.2. However, the noticeably recorded excrescence in the Mulliken bond electronegativity, left-down representation in Figure 3.14, is founded only in the environment coupling range in which the electronic number approaches to one electron in the $|0\rangle$ state and to two electrons in the excited $|g\rangle$ state, respectively, see the top and the middle pictures of Figure 3.3. This excrescence gives the information of the charge accumulation in the bond and shall be associated with the covalent binding (2 electrons *per* bond) case. The absolute arithmetic average electronegativity situation, see the left-upper hyper-surface in Figure 3.14, as well as the geometric electronegativity averages in Figure 3.13 do not pose such a feature.

Regarding the exchanged energies between the involved electronic vibrationally states in the binding, there is remarked a fine concordance of the energetic scale, converted from the atomic units into eV, with the observed values, $1 \div 3$ eV, for that the concerned vibrational bonding takes place. [14]

3.4 OUTLOOK

The present chapter was dedicated to the implementations of the density functionals introduced and deduced in the Chapter 2, for the phenomenological case of the molecular electronic systems on the fundamental and excited vibrationally states.

The markovian methodology to compute the quantum statistic electronic density, in considered coupling conditions, have been employed from the Feynman-Kleinert path integral formalism, and the basic working formulas were displayed.

Within markovian limit, see equation 2.240, there was realized that the Feynman-Kleinert approximation of the effective potential, see equation 2.162, fully recovers the smeared out potential 2.163, making in such the whole formalism to work in relation with applied external potential, in an exclusively manner. Such a treatment is meaningfully supported by the *DFT* theorems, by assuming the external potential as the main input for the electronic structure (density) description.

Such an approach was used to calculate the partitions functions, see equations 3.3, and their electronic densities, see equations 3.4. These last ones were considered in such to display the explicit quantum character and, simultaneously, to do not integrate necessarily always to unity, see equations 3.5 and Figure 3.3, for instance. This way, is possible to treat the entire electronic system as an open one, to whom the changing in charge and energy is based upon the thermal coupling with an environmental reservoir as well as on its coupled vibrational excitations.

The hyper-surfaces of the calculated and represented density functionals have shown a coherent phenomenology of the analysis and interpretation for the considered harmonic – anharmonic molecular electronic prototype systems within the vibrationally and environment couplings.

As a global view, a general mathematical and physical framework for the interpretation of the chemical reactivity was constructed, throughout the introduced reactivity descriptors: the chemical action, the rate reaction, the chemical field and its period, the absolute and Mulliken electronegativities and their hardnesses companions, the reduced total energy and the partial Hohenberg-Kohn functionals, all of them being calculated and represented on the bi-variance coupling ranges to whom the involved electronic systems are reactively subjected.

Therefore, a new premise to characterize the chemical reactivity by means of the quantum statistics of the coupled multi-electronic systems is in such formulated.

CHAPTER 4.

CONCLUSIONS AND PERSPECTIVES

And if from the eternal ideas you are one...

Giacomo Leopardi – To your women, 45

4.1 INTRODUCTION

The present study approaches a fundamental theoretical direction in order to describe the chemical reactivity through the chemical descriptors and the reactivity indices inferred within the density functional theory framework.

In this context, there are introduced and phenomenological evaluated the density functionals, with a major role in the chemical-physical structure and reactivity. The characterization of the many-electronic systems is therefore made, by means of their capacity to participate in interactions and to the structure transformation processes through the charge and energy exchanges.

The central equations used in determinations of the present work, are the fundamental equations 2.72 and 2.76 of the total energy (E) and of the electronegativity (χ) transformations, respectively for an N-electronic open system, able to exchange charge (dN) and potential (dV) with the environment:

$$dE = -\chi\, dN + \int \rho(x)dV(x)dx\,,$$

$$(4.1)$$

$$-d\chi = 2\eta dN + \int f(x)dV(x)dx$$

$$(4.2)$$

being x the position vector. In these expressions $\rho(x)$ represents the uni-electronic density satisfying the N-normalization condition, $N = \int \rho(x)dx$, whereas $\eta = -1/2(\partial\chi/\partial N)_V$ and $f(x) = -(\partial\chi/\partial V)_N$ represent the chemical hardness and the Fukui or the frontier function for the considerate electronic system, respectively.

The chemical descriptors introduced in the present work, and next listed, show a general quantum statistic character allowing a wide range of electronic density implementations associated to the aimed electronic states.

4.2 ORIGINAL CONTRIBUTIONS

In the Chapter 2, by considering the equations 4.1 and 4.2 for transformations of the electronic systems from a ground state to another new reactivity descriptors are introduced, as in summary here follow.

(i)   The chemical action, see equation 2.53, and its principle 2.165 and 2.170:

$$C_A = \int \rho(x)V(x)dx \,;\, \delta C_A = 0 .$$

(4.3)

(ii)   The chemical field, see equation 2.183:

$$\omega_C = \frac{C_A}{\hbar\beta \, V(x)}$$

(4.4)

with the associated chemical field period:

$$T_C = \frac{2\pi}{\omega_C}$$

(4.5)

being β and $\hbar$ the electronic system's inverse thermal energy and the reduced Planck constant, respectively.

(iii)   New rate reaction functional, see equation 2.191:

$$\Gamma[\rho] = \frac{\left[\mu - \frac{1}{\sqrt{a}}\arctan\left(\frac{N}{\sqrt{a}}\right)\right](N^2 + a) - b}{\hbar\beta \, N C_A}$$

(4.6)

where,

$$a = \int_{-\infty}^{+\infty} L(x)dx \,,\, b = \int_{-\infty}^{+\infty} L(x)V(x)dx \,,\, L(x) = -\frac{\nabla\rho(x)}{\nabla V(x)} \quad (4.7)$$

according with the expressions 2.187 and 2.101, respectively, and where the chemical potential μ assumes the general physically formulation, see equation 2.194:

$$\mu = \frac{1}{N\beta} \ln Z - \frac{1}{N} \int \nabla V(x) dx$$

(4.8)

being Z the partition function of the considerate electronic sample.

(iv) Worth noting that for the derivation of the rate reaction 4.6 a new density functional for the absolute electronegativity was used, see equation 2.186, namely:

$$\chi(N) = -\frac{1}{\sqrt{a}} \arctan\left(\frac{N}{\sqrt{a}}\right) - \frac{b}{a+N^2} - NC_A \frac{1}{a+N^2}.$$

(4.9)

(v) Further, the absolute electronegativity 4.9 can be next employed to produce the corresponding Mulliken electronegativity density functional, see equation 2.221, under the form:

$$\chi_M(N) = \frac{b+N-1}{2\sqrt{a}} \arctan\left(\frac{N-1}{\sqrt{a}}\right) - \frac{b+N+1}{2\sqrt{a}} \arctan\left(\frac{N+1}{\sqrt{a}}\right)$$

$$+ \frac{C_A-1}{4} \ln\left[\frac{a+(N-1)^2}{a+(N+1)^2}\right].$$

(4.10)

This new definition of the Mulliken electronegativity was particularized for the atomic systems by using the pseudopotentials as the input information and the path integral Feynman-Kleinert within the Markovian specialization as the computational density implementation. The results, presented in the Section 2.3.9 and the Figure 2.4 are satisfactory compared with the previous data. More, a recent atomic radii scale and related size properties were provided on base of this 4.10 electronegativity formulation. [168]

(vi)   Then, by using the landmark definition of the density functional electronegativity concept, as equation 2.6 reveals, the general Hohenberg-Kohn density functional can be introduced through the relations 2.204,

$$F_{HK} = -\int_0^N \chi(N)dN$$

(4.11)

to become the working partial Hohenberg-Kohn one, see equation 2.214,

$$F_{HK}^P = \frac{b+N}{\sqrt{a}} \arctan\left(\frac{N}{\sqrt{a}}\right) - \frac{1}{2}\ln\left(\frac{a+N^2}{a}\right)$$

$$\times \left[1 + \frac{2\dfrac{b+1}{\sqrt{a}}\arctan\left(\dfrac{1}{\sqrt{a}}\right) - \ln\left(\dfrac{a+1}{a}\right)}{1 + \ln\left(\dfrac{a+1}{a}\right)}\right]$$

(4.12)

from which, its universal formulation can be achieved by considering the potentials' successive solution of the Poisson equation,

$$\nabla^2 V(x) = -4\pi\rho(x)$$

(4.13)

back into the components 4.7.

(vii)   Moreover, since the Hohenberg-Kohn functional 4.11 is combined with the chemical action 4.3, one gets the total energy of the system in terms of the electronegativity, see equation 2.203,

$$E = -\int_0^N \chi(N)dN + C_A$$

(4.14)

with its reduced working formulation, given as in 2.212:

$$-E^R = T = -\frac{b+N}{\sqrt{a}}\arctan\left(\frac{N}{\sqrt{a}}\right) + \frac{1}{2}\ln\left(\frac{a+N^2}{a}\right)$$

$$-\frac{1+\frac{1}{2}\ln\left(\dfrac{a+N^2}{a}\right)}{1+\ln\left(\dfrac{a+1}{a}\right)}\left[\ln\left(\frac{a+1}{a}\right) - 2\frac{b+1}{\sqrt{a}}\arctan\left(\frac{1}{\sqrt{a}}\right)\right].$$

$$(4.15)$$

(viii) Also new chemical hardness formulations, as density functionals, were inferred from the absolute and the Mulliken electronegativities 4.9 and 4.10, respectively, by employing the relationship 2.9, to produce the results 2.229:

$$\eta^\chi = \frac{1}{2(a+N^2)} + \frac{(a-N^2)C_A - 2Nb}{2(a+N^2)^2},$$

$$\eta^{\chi_M} = \frac{N^4 + 2N^2(a-1) + (a+1)^2}{4\sqrt{a}\left[a+(N-1)^2\right]\left[a+(N+1)^2\right]}\left[\arctan\left(\frac{N+1}{\sqrt{a}}\right) - \arctan\left(\frac{N-1}{\sqrt{a}}\right)\right]$$

$$+\frac{(1+a-N^2)C_A - 2Nb}{2\left[a+(N-1)^2\right]\left[a+(N+1)^2\right]}.$$

$$(4.16)$$

All these introduced and deduced chemical descriptors as the density functionals can be certainly particularized once a given potential and a suitable strategy for the electronic density computation apply.

However, in Chapter 3 were calculated, represented and phenomenologically interpreted the above chemical descriptors, for a generic many-electronic system, governed by a prototype anharmonic potential, subjected to the thermal markovian coupling with the environment.

(ix) The Chapter 3 ends with the introduction of another electronegativity related concept, namely the so called Parr electronegativity, see definition 3.9,

$$\chi_P \equiv \chi_{|g\rangle} - \chi_{|0\rangle}$$

$$(4.17)$$

allowing the prediction of which vibrationally excided state $|g\rangle$ is more reactive respecting with the fundamental $|0\rangle$ one, and therefore, more suited to follow the binding processes. All the present density functionals have produced a coherent analysis and interpretation in accordance with the quantum chemistry conceptual notions and principles.

4.3 USED APPROXIMATIONS

For the computational details the following approximations were used.

The electronic systems were considered evolving within a single (radial) coordinate. For the computation of the electronic density throughout the partition function the path integral quantum statistical picture with the Feynman-Kleinert effective potential approximation was chosen. Further on, the markovian approximation of the ultra-short correlations of the many-electronic system with the environment have been considered,

$$\beta \to 0$$

(4.18)

that has the main merit to cancel the low temperatures quantum fluctuations.

For the applied external molecular potentials, the prototype generalized anharmonic potential:

$$V(x) = k\frac{x^2}{2} + g\frac{x^4}{4}$$

(4.19)

was used, being k and g the force constant and the vibrational coupling parameter, respectively.

Nevertheless, the considered approximations are inherent to a conceptual approach of the aimed reality and does not inferred the phenomenological conflicts neither between them nor respecting the obtained results.

4.4 PERSPECTIVES

The present study opens as well a large area of new perspectives concerning the implementations and of the further possible approaches of the chemical reactivity phenomenology, by means of the fusion between the density functional theory and the path integrals formalism. Some of these future directions of prospecting and analysis will be here specified.

4.4.1 Working Potential Models

(i)   The entire computationally study performed in Chapter 3 can be redone in case of the double-well anharmonic potential:

$$V(x) = -\frac{k}{2}x^2 + \frac{g}{4}x^4$$

(4.20)

or, in the alternative form $V(x) = -(k/2)x^2 + (g/4)x^4 + 1/(4g)$. This kind of potential practically extends the electronic transfer study to the protonic one. For instance, at the crossing of the potential barrier an isomerization reaction is promoted in the adiabatic fundamental electronic state. The tunneling process is encouraged by the very small specific mass of protons in the intra-molecular transfers.

(ii)   More generally, there can be treated the case of the parametric anharmonic potential:

$$V(x) = \frac{a_1}{2}x^2 + \frac{a_2}{4}x^4 .$$

(4.21)

For this potential, the effective dynamic equation has the form: $\dot{x} = -\partial V(x)/\partial x$ which, for the equilibrium condition $\dot{x} = 0$, can be reduced at the equation $a_1 x + a_2 x^3 = 0$. At this point two different cases can be distinguished, depending on the sign of the a_1 parameter: 1) if $a_1>0$, $a_2>0$, the real solution of the equation highlights the event $x_0=0$ as being a stable mesoscopic one; 2) if $a_1<0$, $a_2>0$, the event $x_0=0$ gets an instable character, the stability being separated at the level of the events $x_{1,2} = \pm\sqrt{|a_1|/a_2}$. Can be concluded that once the sign of the harmonic parameter a_1 changes between the types 4.19 and 4.20 of the anharmonic potentials, a broking symmetry arises that leads into a separation of the stable and unstable points-events of reactivity, an effect that emphases on a realization of the "phase" transition from the inter- to the intra-molecular transfer processes. [7]

(iii)   The computational study can be also extended upon the excitonic transfers, subjected to the type potential form:

$$V(x) = \frac{k}{2}x^2 + \frac{g}{4}x^4 + \frac{g}{x} \qquad (4.22)$$

where the electronic transfer is considered only for the excited states.

In all above cases, approaches and approximations, the analytical formulation in the path integral for the partition function and the electronic densities underlines the powerful of a such physical-mathematical tool and characterizes in a suitable manner the transformations of the many-electronic systems open to a large couplings classes.

4.4.2 Analytical Molecular Implementations

(i)   One of the meaningful molecular representations that starts from the atomic information regards the *force* approach. [108] By combining the definition of the applied force as the gradient of the exerted potential with the solution of the Poisson equation 4.13, see also the equation 2.217, there is provided the *atomic force* expression in terms of the electronic density of the concerned atomic quantum shell $|n\rangle$, driven on the sense $\vec{n}$ within molecule:

$$-\nabla V_{|n\rangle}(x) \equiv \vec{F}_{|n\rangle}(x) = \vec{n}\, 4\pi \int_{\infty}^{x} \rho_{|n\rangle}(v)\,dv \ .$$

(4.23)

This approach demands the knowledge of the relative atomic positions in the considered molecule and thus also the bond lengths. This information can be released both from the X-ray diffraction measurements as well as from *ab initio* or other theoretical geometric molecular optimizations. Having these data the *molecular (M) resultant force*, from those atomic, simply becomes:

$$F_M(x) = \sqrt{\left(\sum_n \vec{F}_{|n\rangle}(x)\right)^2} \ .$$

(4.24)

Next, one has to take into account that fact that at equilibrium the formed molecule has the optimal geometry that displays the zero force resultant, as the natural bonding condition:

$$F_M(x) = 0 \Rightarrow x \equiv R_M \ .$$

(4.25)

The solution R_M provided by the equation 4.25 has the role of the relevant (fuzzy) "spherical" radius beyond of which the molecular force 4.24

and its associate molecular potential vanish. Therefore, the working molecular potential and the density, within the force approach, get respectively the expressions:

$$V_M(x) = \int_x^{R_M} 4\pi\ w^2 F_M(w)dw\ ,$$

$$\rho_M(x) = -\frac{1}{4\pi}\nabla V_M(x).$$

(4.26)

Finally, all the density functional chemical descriptors presented in this study can be re-evaluated for any molecular concrete system, since the atomic set of densities and the bonding configuration are known as the input information, as this force molecular analytical model demands.

(ii)   Another way to analytically compute the molecular density is based on the *information* approach. In this case, the minimum missing information (entropy deficiency) principle produces the atom in molecules densities as the pieces of the molecular density that optimally resembles the densities of the corresponding isolated atoms. Within the Kulluback-Leibler (*KL*) approach, [169] the missing information looks like:

$$\Delta S^{KL}\left[\rho_M(x)\big|\rho_0(x)\right] = \int \rho_M(x)\ln\left[\frac{\rho_M(x)}{\rho_0(x)}\right]dx\ ,$$

$$\rho_0(x) = \sum_n \rho_{|n\rangle}(x).$$

(4.27)

Now, having the concrete atomic isolated densities $\rho_{|n\rangle}(x)$, the evaluation of the molecular optimal one requires the assumption of an effective trial density, here proposed, under the molecular parametric $m - q - l$ form:

$$\rho_M(x) := mx^q \exp(-l\,x^2).$$

(4.28)

The introduced parameters m, q and l, in the molecular density form 4.28, are to be determined by fulfilling a set of natural conditions: the normalization condition of the total number of electrons in molecule (N_M),

$$\int \rho_M(x)dx = N_M$$

(4.29)

the global equilibrium condition,

$$\sum_n \int \frac{\delta \Delta S^{KL}\big[\rho_M(x)\big|\rho_0(x)\big]}{\delta \rho_{|n\rangle}(x)}\delta \rho_{|n\rangle}(x)dx = 0$$

(4.30)

and the minimum missing information condition between the atomic resembled and optimal molecular densities, abstracted from 4.27, like the equation:

$$\Delta S^{KL}\big[\rho_M(x)\big|\rho_0(x)\big] = 0 .$$

(4.31)

Once the analytic molecular electronic effective density, expression 4.28, have been prescribed, the associate molecular potential further emerges out from the Poisson equation 4.13, see also expressions 2.217.

This way, all the required inputs are provided to allow the molecular computation of the reactivity indices of this study, as the density functionals, for a large class of molecular systems.

(iii)   Last method to analytical treat the concrete molecular systems regards the assumption of the input trial molecular potential, instead of the previous molecular density. For this case the suitable trial molecular potential states the generalized anharmonic one, the 3.1 type potential:

$$V_{|g\rangle}(x) = \frac{1}{2}k_M x^2 + \frac{1}{4}gx^4 .$$

(4.32)

This time, within such potential approach, all required analytical relations are presented in the Section 3.2 as a three parametric $k_M - g - \beta$ dependency.

With the aim to individuate these parameters for, in principle, any molecular system the appropriate conditions have to be imposed.

For the determination of the molecular force constant k_M is enough to assume the individual atoms as the hydrogenoid ones with the concerned bonding electrons in their atomic quantum valence shells evolving under the core influence, encompassed within effective nuclear charge Z_{eff}. Thus, the quantum Bohr model for the hydrogenoid atoms can be applied to find out their effective force constants. This procedure repeats for each bonding atom with its bonding participant (effective) electron. Next on, the net of molecular bonds can be modeled as the combined atoms in molecule's oscillations bonding with their individual force constants. Performing the respective composed serial-parallel oscillations will result the effective molecular force constant k_M.

Once k_M is determined, the remaining parameters, g and β, can be determined by solving the system consisting of the molecular density condition 4.29 as the 3.5 adapted equation,

$$N_M = 2\sqrt{\frac{2\pi\beta\, k_M}{4k_M + g\beta}} \frac{K_{\frac{1}{4}}\left[\dfrac{\beta\, k_M^2}{8g}\right]}{K_{\frac{1}{4}}\left[\dfrac{\beta\,(4k_M + g\beta)^2}{128g}\right]} \exp\left[-\frac{\beta^2\left(8k_M + g\beta^3\right)}{384}\right]$$

(4.33)

being $K[]$ the modified Bessel functions of the second rank, together with the maximum chemical hardness condition, see Section 2.2.3 and the relation 2.71, which, for the actual approach, means the differential condition:

$$\left(\frac{\partial\eta}{\partial g}\right)_{\beta,\, k_M} = 0\,.$$

(4.34)

Nevertheless, for the chemical hardness expression its general formulations 2.229 have to be analytically specialized with the expressions given in the Section 3.2. Finally, since all the $k_M - g - \beta$ parameters have been identified, any interested reactivity index can be evaluated on the particular studied molecular case.

4.5 FINAL OUTLOOK

The presented study, throughout the introduced and deduced density functionals, contribute in providing a general mathematical-physical framework for the chemical reactivity and the structure transformation processes, in terms of associate density functional descriptors within the conceptual quantum chemistry. Both analytical functionals as well the performed implementations were studied in detail and have been phenomenologically applied and interpreted. This way the physico-chemical "analytical space" for a large class of many-electronic systems analysis was enlarged.

APPENDIX

Nobody with common sense would can deny it...

Plato – The Republic, 9, 580 a

In these sections there will be presented the mathematical details of some expressions used without demonstration in the basic text of the Chapter 2.

A.0 THE POISSON FORMULA FOR SERIES

The Poisson formula permits the transformation of the series $\sum\limits_{m=-\infty}^{+\infty} f(m)$ into another one for which the evaluation is (most of the times) easier.

One will start from the consideration of the "comb" function:

$$S(q) = \sum_{m=-\infty}^{+\infty} \delta(q-m)$$

(A.1)

seen as series of the delta (Dirac) functions, with the representation in the Figure A.1.

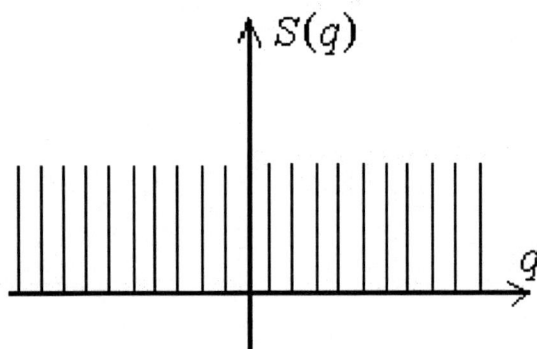

Figure A.1. *The representation of the "comb" function.*

From the definition A.1, and/or from Figure A.1, it can be observed the periodicity property of the "comb" function:

$$S(q+n) = S(q); \; n \in Z$$

(A.2)

that allows to develop the original function in Fourier series:

$$S(q) = \sum_{n=-\infty}^{+\infty} S_n \exp(-i2\pi qn) \; .$$ (A.3)

In the series A.3 the coefficient of the Fourier expansion has the expression:

$$S_n \equiv \int_{-1/2}^{+1/2} dq S(q) \exp(i2\pi nq) = 1$$

(A.4)

where, for getting the last identity of A.4 the expression A.1 as well as the properties of the delta Dirac function have been used. By submitting the result A.4 back in the expression A.3 results also the identity:

$$\sum_{m=-\infty}^{+\infty} \delta(q-m) = \sum_{n=-\infty}^{+\infty} \exp(-i2\pi\, qn).$$

(A.5)

When the terms of the relation A.5 are multiplied with the integral $\int_{-\infty}^{+\infty} dq f(q)$ and the properties of the delta Dirac function counts again, the final Poisson formula for series springs out to be:

$$\sum_{m=-\infty}^{+\infty} f(m) = \sum_{n=-\infty}^{+\infty} \left[\int_{-\infty}^{+\infty} dq f(q) \exp(-i2\pi\, qn) \right].$$

(A.6)

The expression A.6 seems, at the first sight, like a complicated version of the initial series (the left term of the relation A.6), whereas in the concrete applications it proves to be a very useful transformation of the initial series into a convergent one (see also the next Sections).

A.1 THE PERIODICAL PATHS

A method to introduce the quantum statistical path integrals is to consider the periodical $x(\tau)$ paths:

$$x(0) = x(\hbar\beta)$$

(A.7)

with an representation depicted in the Figure A.2. In these conditions, the path integrals can be written by cumulating the integrations on the coordinate axes:

$$\oint D(x(\tau)) \equiv \int_{-\infty}^{+\infty} dx \int_{x(0)=x(\hbar\beta)} Dx(\tau) .$$

(A.8)

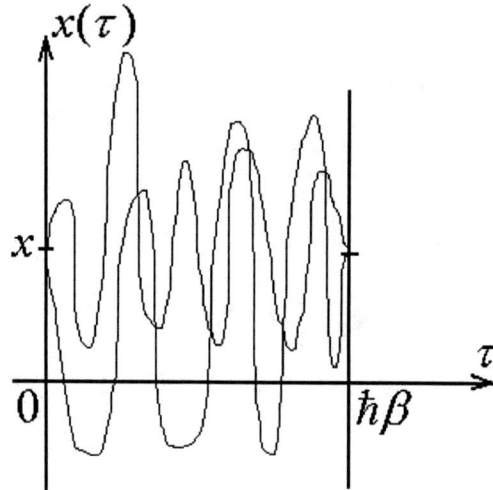

Figure A.2. *The representation of the periodical quantum statistical paths.*

From the periodicity condition A.7, one realizes that the involved paths $x(\tau)$ can be developed into Fourier series:

$$x(\tau) = \sum_{m=-\infty}^{+\infty} x_m \exp(i\omega_m \tau) .$$

(A.9)

Since the particularization of the series A.9 is done for the "times" concerned in the A.7 condition, there follow the expressions:

$$x(0) = \sum_{m=-\infty}^{+\infty} x_m ,$$

$$x(\hbar\beta) = \sum_{m=-\infty}^{+\infty} x_m \exp(i\omega_m \hbar\beta)$$

(A.10)

from where, by equating them according with the condition A.7 the so called Matsubara frequencies result:

$$\omega_m = \frac{2\pi}{\hbar\beta} m .$$

(A.11)

In addition, under the condition that the considered paths should be real:

$$x^*(\tau) = x(\tau) ,$$

(A.12)

$$x^*(\tau) = \sum_{m=-\infty}^{+\infty} x_m^* \exp(-i\omega_m\tau) \overset{m->-m}{=} \sum_{m=-\infty}^{+\infty} x_{-m}^* \exp(i\omega_m\tau)$$

(A.13)

the relations among the coefficients of the periodical paths expansion are next released:

$$x_m = x_{-m}^* ,$$

(A.14)

$$x_0 = \mathrm{Re}(x_0) .$$

(A.15)

These properties permit to rewrite the integral A.8 as the decomposed form on the real and imaginary coefficients of the periodical paths,

$$\oint D(x(\tau)) \equiv \left[N_0 \int_{-\infty}^{+\infty} dx_0 \right] \left[\prod_{m=1}^{\infty} N_m \int_{-\infty}^{+\infty} d\,\mathrm{Re}(x_m) \int_{-\infty}^{+\infty} d\,\mathrm{Im}(x_m) \right]$$

(A.16)

and where, the integration constants N_0 and N_m are aimed to satisfy the normalization exigencies of the quantum statistical path integrals and will be here analytically determined, see the form A.43.

A.2 THE GENERALIZED RIEMANN SERIES

An important application of the Poisson formula for series, the relation A.6, regards the calculation of the series:

$$f(\Omega) = \sum_{m=1}^{\infty} \frac{1}{\omega_m^2 + \Omega^2}$$

(A.17)

seen as a generalization of the well known Riemann series:

$$f_R = \sum_{m=1}^{\infty} \frac{1}{m^2} = \frac{\pi^2}{6}.$$

(A.18)

In order to calculate the series A.17, firstly, it will be reconsidered under the form:

$$f(\Omega) = \sum_{m=1}^{\infty} \frac{1}{\omega_m^2 + \Omega^2} = \frac{1}{2}\left(\sum_{m=-\infty}^{\infty} \frac{1}{\omega_m^2 + \Omega^2} - \frac{1}{\Omega^2} \right)$$

(A.19)

being the remaining problem the evaluation of the series:

$$F(\Omega) = \sum_{m=-\infty}^{\infty} \frac{1}{\omega_m^2 + \Omega^2}$$

(A.20)

with ω_m identified as the Matsubara frequencies A.11.

If one applies the Poisson formula A.6 on the series A.20 and replaces also the Matsubara frequencies A.11, will get the equivalent form:

$$F(\Omega) = \frac{\hbar^2 \beta^2}{4\pi^2} \sum_{n=-\infty}^{\infty} \left[\int_{-\infty}^{+\infty} dq \frac{1}{q^2 + \alpha^2} \exp(-i2\pi q n) \right],$$

$$\alpha^2 \equiv \frac{\hbar^2 \beta^2 \Omega^2}{4\pi^2}.$$

(A.21)

There can be observed that the calculus of the series in A.21 firstly demands the evaluation of the raised integral. The function under the integral presents the poles $q = \pm i\alpha$, from where, appears the necessity of its calculation in the complex plan.

Due to the fact that:

$$\left| \exp\left[-i2\pi \, (\operatorname{Re} q + i \operatorname{Im} q)n \right] \right| = \exp(2\pi \, n \operatorname{Im} q)$$

(A.22)

in order to assure the convergence of the integral in A.21, two contour complex path cases can be considered, namely:

$$n \geq 0 \Rightarrow \operatorname{Im} q < 0 \Rightarrow \text{path } II,$$

$$n < 0 \Rightarrow \operatorname{Im} q > 0 \Rightarrow \text{path } I$$

being their integration outlines presented in the Figure A.3 below.

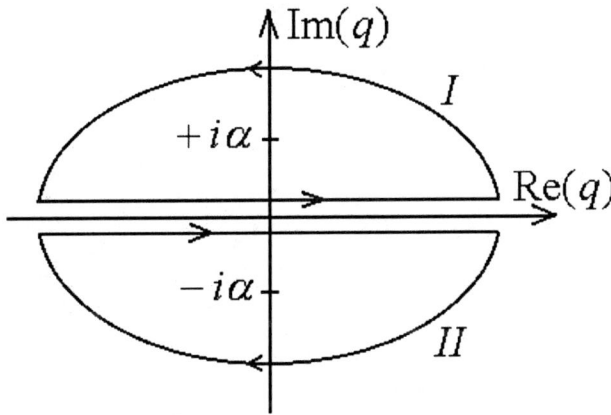

Figure A.3. *The integration contours around the poles* $q = \pm i\alpha$.

Therefore, the integral in A.21 will be written as a sum of two integrals corresponding to the actual complex outlines.

The path I is considered as being that one with the mathematical sense, whereas the path II is assigned to the anti-mathematical one that is equivalent with the (-1) sign in front of the corresponding integral. For the effective integration the residue theorem applies with the prescription:

$$\oint_{C(z_0)} f(z)dz = 2\pi \, i \operatorname*{Rez}_{z=z_0} f(z) = 2\pi \, i \lim_{z \to z_0} (z - z_0)f(z).$$

(A.23)

For instance, for the outline *II* of the Figure A.3, the complex integration of the integral in A.21 successively writes:

$$\int_{-\infty}^{+\infty} dq \frac{1}{q^2 + \alpha^2} \exp(-i2\pi\, qn) = (-1) \oint_{C(II)} dq \frac{1}{q^2 + \alpha^2} \exp(-i2\pi\, qn)$$

$$= -2\pi\, i \operatorname*{Rez}_{q=-i\alpha} \frac{\exp(-i2\pi\, qn)}{q^2 + \alpha^2}$$

$$= -2\pi\, i \lim_{\substack{q=-i\alpha \\ \alpha = \frac{\hbar\beta\Omega}{2\pi}}} (q + i\alpha) \frac{\exp(-i\,2\pi\, qn)}{(q + i\alpha)(q - i\alpha)}$$

$$= \frac{2\pi}{\hbar\beta\,\Omega}^2 \exp(-\hbar\beta\,\Omega n)$$

$$\tag{A.24}$$

from where, there can be observed that the integration's result is a real quantity. The output A.24 will be the same also for the complex integral on the outline *II* in Figure A.3. In these conditions, expression A.21 becomes:

$$F(\Omega) = \frac{\hbar^2\beta}{4\pi}^2 \left\{ \sum_{n=0}^{\infty} \left[\int_{-\infty}^{+\infty} dq \frac{\exp(-i2\pi\, qn)}{q^2 + \alpha^2} \right] + \sum_{n=1}^{\infty} \left[\int_{-\infty}^{+\infty} dq \frac{\exp(i2\pi\, qn)}{q^2 + \alpha^2} \right] \right\}$$

$$= \frac{\hbar^2\beta}{4\pi}^2 \left\{ \frac{2\pi}{\hbar\beta\,\Omega}^2 \sum_{n=0}^{\infty} \exp(-\hbar\beta\,\Omega n) + \frac{2\pi}{\hbar\beta\,\Omega}^2 \sum_{n=1}^{\infty} \exp(-\hbar\beta\,\Omega n) \right\}$$

$$= \frac{\hbar\beta}{2\Omega} \left\{ 2\sum_{n=0}^{\infty} [\exp(-\hbar\beta\,\Omega)]^n - 1 \right\}$$

$$= \frac{\hbar\beta}{2\Omega} \left\{ 2\frac{1}{1 - \exp(-\hbar\beta\,\Omega)} - 1 \right\}$$

$$= \frac{\hbar\beta}{2\Omega} \coth\left(\frac{\hbar\beta\Omega}{2} \right). \tag{A.25}$$

With the result A.25 (alias A.20) back in A.19 the generalized Riemann series is founded as:

$$f(\Omega) = \sum_{m=1}^{\infty} \frac{1}{\Omega^2 + \omega_m^2} = \frac{1}{4} \frac{\hbar\beta}{\Omega} \left[\coth\left(\frac{\hbar\beta\,\Omega}{2}\right) - \frac{2}{\hbar\beta\,\Omega} \right].$$

(A.26)

From the last expression also the series A.18 it can be recovered once the condition $\Omega \to 0$ (of the free motion), the expansion of the hyperbolic cotangent series $\coth z \approx 1/z + z/3$ for $z \to 0$, and the Matsubara frequencies A.11, successively apply.

A.3 THE NORMATION OF THE PERIODIC PATH INTEGRALS

In order to express the normalization constants N_0 and N_m for completion the path integral A.16, there will be considered the referentially partition function of the harmonic oscillator written under the path integral form:

$$Z_\Omega = \oint D(x(\tau)) \exp\left\{ -\frac{1}{\hbar} \int_0^{\hbar\beta} d\tau \left[\frac{m_0}{2} \dot{x}^2(\tau) + \frac{m_0}{2} \Omega^2 x^2(\tau) \right] \right\}$$

(A.27)

knowing that the result of this integration has to has the form:

$$Z_\Omega = \sum_{n=0}^{\infty} \exp\left[-\beta\,\hbar\Omega\left(n + \frac{1}{2} \right) \right]$$

$$= \exp\left(-\frac{\hbar\beta\,\Omega}{2} \right) \sum_{n=0}^{\infty} [\exp(-\beta\,\hbar\Omega)]^n$$

$$= \exp\left(-\frac{\hbar\beta\,\Omega}{2} \right) \frac{1}{1 - \exp(-\hbar\beta\,\Omega)}$$

$$= \frac{1}{2\sinh(\hbar\beta\,\Omega/2)}.$$

(A.28)

With the aim to equate the two above, A.27 and A.28, partition function forms, one can evaluate the temporal integrals in A.27 by using the Fourier decomposition A.9 together with the Matsubara frequencies A.11. This way, successively, there are obtained the equivalent expressions:

$$\frac{m_0}{2}\Omega^2 \int_0^{\hbar\beta} d\tau\, x^2(\tau) \stackrel{(A.13)}{=} \frac{m_0}{2}\Omega^2 \int_0^{\hbar\beta} d\tau \left[\sum_{m=-\infty}^{+\infty} x_m \exp(i\omega_m\tau)\right]\left[\sum_{m'=-\infty}^{+\infty} x_{m'} \exp(i\omega_m\tau)\right]$$

$$= \frac{m_0}{2}\Omega^2 \sum_{m=-\infty}^{+\infty}\sum_{m'=-\infty}^{+\infty} x_m x_{m'} \int_0^{\hbar\beta} d\tau\, \exp[i(\omega_m+\omega_{m'})\tau]$$

$$= \frac{m_0}{2}\Omega^2 \sum_{m=-\infty}^{+\infty}\sum_{m'=-\infty}^{+\infty} x_m x_{m'} \hbar\beta\delta_{m+m',0}$$

$$= \frac{m_0}{2}\Omega^2\hbar\beta \sum_{m=-\infty}^{+\infty} x_m x_{-m}$$

$$= \frac{m_0}{2}\Omega^2\hbar\beta \sum_{m=-\infty}^{+\infty} x_m x_m^*$$

$$= \frac{m_0}{2}\Omega^2\hbar\beta \left\{x_0^2 + 2\sum_{m=1}^{+\infty}\left[(\operatorname{Re}x_m)^2 + (\operatorname{Im}x_m)^2\right]\right\}$$

$$(A.29)$$

and, respectively:

$$\frac{m_0}{2} \int_0^{\hbar\beta} d\tau\, \dot{x}^2(\tau) \stackrel{(A.13)}{=} \frac{m_0}{2} \int_0^{\hbar\beta} d\tau \left[\sum_{m=-\infty}^{+\infty} x_m\omega_m \exp(i\omega_m\tau)\right]\left[\sum_{m'=-\infty}^{+\infty} x_{m'}\omega_{m'} \exp(i\omega_m\tau)\right]$$

$$= \frac{m_0}{2} \sum_{m=-\infty}^{+\infty}\sum_{m'=-\infty}^{+\infty} \omega_m\omega_{m'} x_m x_{m'} \int_0^{\hbar\beta} d\tau\, \exp[i(\omega_m+\omega_{m'})\tau]$$

$$= \frac{m_0}{2} \sum_{m=-\infty}^{+\infty}\sum_{m'=-\infty}^{+\infty} \omega_m\omega_{m'} x_m x_{m'} \hbar\beta\delta_{m+m',0}$$

$$= \frac{m_0}{2} \hbar\beta \sum_{m=-\infty}^{+\infty} \omega_m \omega_{-m} x_m x_{-m}$$

$$= \frac{m_0}{2} \hbar\beta \sum_{m=-\infty}^{+\infty} \omega_m \omega_m^* x_m x_m^*$$

$$= m_0 \hbar\beta \sum_{m=1}^{+\infty} \omega_m^2 \left[(\mathrm{Re}\, x_m)^2 + (\mathrm{Im}\, x_m)^2 \right].$$

(A.30)

Then, if one replaces the expressions A.16, A.29, and A.30 in the A.27 integral, the partition function of the harmonic oscillator equally becomes:

$$Z_\Omega = \left\{ N_0 \int_{-\infty}^{+\infty} dx_0 \prod_{m=1}^{\infty} \left[N_m \int_{-\infty}^{+\infty} d\,\mathrm{Re}(x_m) \int_{-\infty}^{+\infty} d\,\mathrm{Im}(x_m) \right] \right\}$$

$$\times \exp\left\{ -m_0\beta \sum_{m=1}^{+\infty} \omega_m^2 [(\mathrm{Re}\, x_m)^2 + (\mathrm{Im}\, x_m)^2] - \frac{m_0}{2}\Omega^2\beta\, x_0^2 \right.$$

$$\left. -m_0\Omega^2\beta \sum_{m=1}^{+\infty} [(\mathrm{Re}\, x_m)^2 + (\mathrm{Im}\, x_m)^2] \right\}$$

$$= \left[N_0 \int_{-\infty}^{+\infty} dx_0 \exp\left(-\frac{m_0}{2}\Omega^2\beta\, x_0^2 \right) \right]$$

$$\times \prod_{m=1}^{\infty} N_m \left\{ \int_{-\infty}^{+\infty} d\,\mathrm{Re}(x_m) \exp\left[-m_0\beta \sum_{m=1}^{\infty} (\omega_m^2 + \Omega^2)(\mathrm{Re}\, x_m)^2 \right] \right\}^2$$

$$= N_0 \sqrt{\frac{2\pi}{m_0\beta\,\Omega^2}} \prod_{m=1}^{\infty} N_m \frac{\pi}{m_0\beta\,(\omega_m^2 + \Omega^2)}.$$

(A.31)

If for N_m one chooses the form $N_m = C\omega_m^2$, the partition function A.31 can be further written in a short-handed manner:

$$Z_\Omega = N_0 \frac{C}{\Omega} \sqrt{\frac{2\pi}{m_0\beta}} \frac{\pi}{m_0\beta} h(\Omega)$$

(A.32)

where the additional function was introduced:

$$h(\Omega) = \prod_{m=1}^{\infty} \frac{\omega_m^2}{\omega_m^2 + \Omega^2}.$$

(A.33)

Now, the series A.33 can be further employed as:

$$g(\Omega) = \ln h(\Omega) = \sum_{m=1}^{\infty} \left[\ln \omega_m^2 - \ln(\omega_m^2 + \Omega^2)\right] = \sum_{m=1}^{\infty} \ln\left[\frac{\omega_m^2}{\omega_m^2 + \Omega^2}\right]$$

(A.34)

by which, its first derivative respecting to the parameter Ω provides that:

$$g'(\Omega) = \sum_{m=1}^{\infty} \frac{\omega_m^2 + \Omega^2}{\omega_m^2} \frac{-\omega_m^2}{(\omega_m^2 + \Omega^2)^2} 2\Omega$$

$$= -2\Omega \sum_{m=1}^{\infty} \frac{1}{\omega_m^2 + \Omega^2}$$

$$\overset{(A.26)}{=} -\frac{\hbar\beta}{2}\left[\coth\left(\frac{\hbar\beta\,\Omega}{2}\right) - \frac{2}{\hbar\beta\,\Omega}\right].$$

(A.35)

Then, integrating back the relation A.35,

$$g(\Omega) = g(0) + \int_0^\Omega d\tilde{\Omega}\, g'(\tilde{\Omega})$$

(A.36)

and considering the variable change $z = \hbar\beta\Omega/2$, one successively gets:

$$g(\Omega) = - \int_0^{\frac{\hbar\beta\,\Omega}{2}} dz \left(\coth z - \frac{1}{z} \right)$$

$$= -[\ln(\sinh z) - \ln z]_0^{\frac{\hbar\beta\,\Omega}{2}}$$

$$= -\left[\ln\left(\frac{\sinh z}{z} \right) \right]_0^{\frac{\hbar\beta\,\Omega}{2}}$$

$$= \ln\left[\frac{\hbar\beta\,\Omega}{2\sinh(\hbar\beta\,\Omega/2)} \right].$$

(A.37)

Comparing now the expressions A.34 and A.37, springs out the value of the series A.33 and finally of the partition function A.32 as:

$$Z_\Omega = N_0 C \frac{1}{\Omega} \sqrt{\frac{2\pi}{m_0\beta}} \frac{\pi}{m_0\beta} \frac{\hbar\beta\,\Omega}{2\sinh(\hbar\beta\,\Omega/2)}.$$

(A.38)

Further comparing, this time, the partition functions A.28 and A.38, both of them expressing the partition function of the harmonic oscillator, the relationship among the integration constants yields to be:

$$1 = N_0 C \sqrt{\frac{2\pi}{m_0\beta}} \frac{\pi\hbar}{m_0}.$$

(A.39)

With the aim of solving A.39, there is taking into account of the conventional form for expressing the partition function through out the thermal length $\lambda_{th} = \sqrt{2\pi\,\hbar^2\beta\,/m_0}$, being therefore chosen the specialization:

$$N_0 = \frac{1}{\lambda_{th}} = \frac{1}{\sqrt{2\pi\,\hbar^2\beta\,/m_0}}.$$

(A.40)

With the choice A.40 for N_0 combined with the relation A.39 the constant C is founded like:

$$C = \frac{\beta \, m_0}{\pi}$$

(A.41)

providing also the result:

$$N_m = C\omega_m^2 = \frac{\beta \, m_0 \omega_m^2}{\pi} \, .$$

(A.42)

Finally, with the help of the relations A.40 and A.42 the path integral A.16 can be rewritten as the normalized form:

$$\oint D(x(\tau)) \equiv \left[\int_{-\infty}^{+\infty} \frac{dx_0}{\sqrt{2\pi \, \hbar^2 \beta / m_0}} \right] \left[\prod_{m=1}^{\infty} \int_{-\infty}^{+\infty} \int_{-\infty}^{+\infty} \frac{d \operatorname{Re} x_m \, d \operatorname{Im} x_m}{\pi / (m_0 \beta \, \omega_m^2)} \right] \, .$$

(A.43)

NOTES AND REFERENCES

Bless who can know the reasons

Virgil – The Georgians, 2, 490

1.   Atkins, P. W. (1995-Romanian Edition). *The Periodic Kingdom*, Humanitas: Bucharest.

2.   Bohm, D. (1995-Romanian Edition) *Wholeness and the Implicate Order*, Humanitas: Bucharest.

3.   Dawkins, R. (1995-Romanian Edition). *River out of Eden*, Humanitas: Bucharest.

4.   Einstein, A. (1992-Romanian Edition). *Mein Weltbild*, Humanitas: Bucharest.

5.   Gammaitoni, L.; Haenggi, P.; Jung, P.; Marchesoni, F. (1998). Stochastic resonance, *Rev. Mod. Phys.*, **70**, 224-287.

6.   Hänggi, P.; Talkner, P.; Borkovec, M. (1990). Reaction-rate theory: fifty years after Kramers, *Rev. Mod. Phys.*, **62**, 225-341.

7.   Haken, H. (1978). *Synergetics*, Springer-Verlag: Heidelberg

8.   Haken, H. (1987). *Advanced Synergetics*, Springer-Verlag: Heidelberg.

9.   Haken, H. (1988). *Information and Self-Organization*, Springer-Verlag: Heidelberg.

10. Prigogine, I.; Stengers, I. (1997-Romanian Edition). *Entre le Temps et L'éternité*, Humanitas: Bucharest.

11. Hawking, S. (1995-Romanian Edition). *A Brief History of Time*, Humanitas: Bucharest.

12. Einstein, A.; Podolsky, B.; Rosen, N. (1935). Can quantum-mechanical description of physical reality be considered complete?, *Phys. Rev.*, **47**: 777-780.

13. Parr, R. G. (1972). *The Quantum Theory of Molecular Electronic Structure*, W. A. Benjamin, Inc.: Reading, Massachusetts.

14. May, V.; Kühn, O. (2001). *Charge and Energy Transfer Dynamics in Molecular Systems*, Wiley-VCH: Berlin.

15. Feynman, R. P.; Hibbs, A. R. (1965). *Quantum Mechanics and Path Integrals*, McGraw-Hill: New York.

16. Hohenberg, P.; Kohn, W. (1964). Inhomogeneous electronic gas, *Phys. Rev.*, **136**: 864-871.

17. Kohn, W.; Sham, L. J. (1965). Self-consistent equations including exchange and correlation effects, *Phys. Rev.*, **140**: 1133-1138.

18. Parr, R. G.; Yang, W. (1989). *Density Functional Theory of Atoms and Molecules*, Oxford University Press: New York.

19. Dreizler, R. M.; Gross, E. K. U. (1990). *Density Functional Theory*, Springer Verlag: Heidelberg.

20. Kohn, W.; Becke, A. D.; Parr, R. G. (1996). Density functional theory of electronic structure, *J. Phys. Chem.*, **100**: 12974-12980.

21. Mermin, N. D. (1965). Thermal properties of the inhomogeneous electron gas, *Phys. Rev.*, **137**: A1441-A1443.

22. Runge, E.; Gross, E. K. U. (1984). Density functional theory for time-dependent systems, *Phys. Rev. Lett.*, **52**: 997-1000.

23.   Neal, H. L. (1998). Density functional theory of one-dimension two-particle systems, *Am. J. Phys.*, **66**: 512-516.

24.   Feynman, R. P.; Kleinert, H. (1986). Effective classical partition function, *Phys. Rev. A*, **34**: 5080-5084.

25.   Kleinert, H. (2001). *Path Integrals in Quantum Mechanics, Statistics and Polymer* Physics, 3rd edition, World Scientific: Singapore. Note: electronic edition available at address: http://www.physik.fu-berlin.de/~kleinert/b3.

26.   Matthews, P. M.; Saphiro, I. I.; Falkoff, D. L. (1960). Stochastic equations for non-equilibrium processes, *Phys. Rev.*, **120**: 1-16.

27.   Pawula, R. F. (1967). Approximation of the linear Boltzmann equation by the Fokker-Planck equation, *Phys. Rev.*, **162**: 186-188.

28.   Risken, H. (1984). *The Fokker-Planck Equation*, Springer-Verlag: Heidelberg.

29.   van Kampen, N. G. (1987). *Stochastic Processes in Physics and Chemistry*, North-Holland.

30.   Gardiner, C. W. (1994). *Handbook of Stochastic Methods*, Springer-Verlag: Heidelberg.

31.   Weiss, U. (1999). *Quantum Dissipative Systems*, 2nd edition, World Scientific: Singapore.

32.   Grăvilă, P.; Putz, M. V.; Grăvilă, C. (1998). Theoretical determination of a phase transition at the adsorption of Na on the Si(001) surface, *Annals of West University of Timisoara, Series of Chemistry*, **7**: 91-94.

33.   Putz, M. V.; Grăvilă, P. (1998). The photo - dissociation of HCl in rare gases matrices, *Annals of West University of Timisoara, Series of Chemistry*, **7**: 95-101.

34.   Putz, M. V. (1998). Electron-phonon coupled dynamics in condensed rare gases, *Annals of West University of Timisoara, Series of Chemistry*, **7**: 131-136.

35.   Putz, M. V.; Chiriac, A. (1998). The matrix density approach of HCl dissociation in rare gases matrices, *Annals of West University of Timisoara, Series of Chemistry*, **7**: 125-130.

36.   Putz, M. V.; Berkoff, V.; Schwentner, N. (1999). The experiment of dissociation of HCl in Xe condensed. The harpoon mechanism, *Scientific Bull. of Polytechnic University of Timisoara, Series of Mathematical-Physics*, **44**: 47-52.

37.   Putz, M. V.; Schwentner, N. (1999). Chemical reactions controlled by lasers (in Romanian), *Medico-Chirurgical Experimental Researches*, **4**: 359-362.

38.   Putz, M. V.; Chiriac, A.; Mracec, M. (2001). Models of photo-mobility selectivity in rare gases matrices, *Rev. Roum. Chimie*, **46**: 153-160.

39. Kleinert, H.; Pelster, A.; Putz, M. V. (2002). Variational perturbation theory for Markov processes, *Phys. Rev. E*, **65**: 66128/1-7. Electronic print: cond-mat/0202378

40. Putz, M. V. (2001). Variational perturbation theory for Markov processes, oral contribution at *International Conference in Mathematics-FILOMAT*, Mathematical Physics Section, Nis-Yugoslavia, August 26-30, URL (WWW) site: http://www.pmf.ni.ac.yu/filomat2001/mathphys.htm , Book of Abstracts: 45.

41. Kleinert, H.; Pelster, A.; Putz, M. V.; Dreger, J.; Hamprecht, B; (2002). Variational perturbation theory for Markov processes, poster contribution at *Deutsche Physikalische Gesellschaft e. V. (DPG)*, Section of Nonlinear Stochastically Systems, Regensburg-Germany, March 11-15, E-Abstracts: http://dpg.rz.uni-ulm.de/archive/2002/dy_21.html

42. Kleinert, H.; Pelster, A.; Putz, M. V.; Dreger, J.; Hamprecht, B; (2002). Variational calculation of conditional probability densities, contribution communicated by Pelster, A. at *The 7th International Conference in Path Integrals, from Quarks to Galaxies*, Anwerpen-Belgium, May 27-31, Book of Abstracts: 56.

43. Putz, M. V.; Chiriac, A. (1999). The chemical reactivity field. I. The density functionals descriptors, *Annals of West University of Timisoara, Series of Chemistry*, **8**: 173-178.

44. Putz, M. V.; Chiriac, A. (1999). The chemical reactivity field. II. The thermo-chemical potentials, *Annals of West University of Timisoara, Series of Chemistry*, **8**: 179-184.

45. Putz, M. V.; Chiriac, A. (1999). The chemical reactivity field. III. The reactivity indices, *Annals of West University of Timisoara, Series of Chemistry*, **8**: 185-188.

46. Putz, M. V. (1999). Statistische quantische electronegativität, *Scientific Bull. of Polytechnic University of Timisoara, Series of Mathematical - Physics*, **44**: 87-91.

47. Putz, M. V. (2000). The quantum statistics of the chemical reactivity. Part I, *Scientific Bull. of Polytechnic University of Timisoara, Series of Mathematical-Physics*, **45**: 62-67.

48. Putz, M. V.; Mracec, M.; Chiriac, A. (2000), Physical bases of the chemical reactivity (in Romanian), *Series of Monographs in Chemistry of West University of Timisoara*, **30**.

49. Putz, M. V.; Chiriac, A.; Mracec, M. (2001). Foundations for a theory of the chemical field. I. General aspects, *Rev. Roum. Chimie*, **46**: 387-395.

50. Putz, M. V.; Chiriac, A.; Mracec, M. (2001). Foundations for a theory of the chemical field. II. The chemical action, *Rev. Roum. Chimie*, **46**: 1175-1181.

51. Putz, M. V.; Chiriac, A.; Mracec, M. (2002). Foundations for a theory of the chemical field. III. The integrated electronegativity, *Rev. Roum. Chimie*, **47**: 201-206.

52. Putz, M. V.; Chiriac, A.; Mracec, M. (2002). Foundations for a theory of the chemical field. IV. The chemical field, *Rev. Roum. Chimie*, to be published.

53. Putz, M. V.; Russo, N. (2002). A new electronegativity scale from path integral formulation, oral contribution at *The 7th International Conference in Path Integrals-from Quarks to Galaxies*, Anwerpen-Belgium, May 27-31, Book of Abstracts: 58.

54. Putz, M. V.; Sicilia, E.; Russo, N. (2002). New atomic radii scale from analytic electronegativities, poster contribution at NATO Advanced Study Institute (ASI) on *Metal-ligand interactions in molecular-, nano-, micro-, and macro-systems in complex environments*, Cetraro-Italy, September 1-12, Book of Abstracts.

55. Putz, M. V. (2002). New electronegativity formulation in density functional theory and its use for atomic properties, oral contribution at NATO Advanced Study Institute (ASI) on *Metal-Ligand interactions in molecular-, nano-, micro-, and macro-systems in complex environments*, Cetraro-Italy, September 1-12; and contribution to the associated Proceeding Book, Russo, N.; Salahub D. R.; Witko, M. (Eds.), (2003), Kluwer Academic Publishers: Dordrecht, in press.

56. Putz, M. V. (2003). Periodic properties from reactivity indices and path integrals approaches, oral contribution at *1st Workshop on The state-of-the-art of Computational Chemistry in the Universities of Calabria and Basilicata*, Arcavacata di Rende-Italy, February 19, URL WWW site: http://unical.it/portale/portalmedia/2003-02/workshop.doc, to be published in the associate Proceeding by the Centre of Excellence MIUR of University of Calabria.

57. Putz, M. V.; Chiriac, A.; Mracec, M. (2000). The thermo-chemical control from the density functional theory, poster presented at *The 2nd International Conference of the Chemical Societies of South-Eastern European Countries*, Halkidiki-Grecia, June 6-9, Book of Abstracts, **1**: 401.

58. Putz, M. V.; Chiriac, A.; Mracec, M. (2000). From density functional theory to the chemical field theory, poster presented at *The 3th European Conference of the Computational Chemistry*, Budapest-Hungary, September 4-8, Book of Abstracts: 72.

59. Putz, M. V.; Chiriac, A.; Mracec, M. (2000). Path integrals approach of the density functional theory, poster presented at *The 3th European Conference of the Computational Chemistry*, Budapest-Hungary, September 4-8, Book of Abstracts: 73.

60. Putz, M. V.; Chiriac, A.; Mracec, M. (2000). The integrated electronegativity in density functional theory, poster contribution at *The 10th Romanian Conference of Physical Chemistry*, Iasi-Romania, September 26-29, Book of Abstracts, **S1**: Po27.

61. Putz, M. V.; Chiriac, A.; Mracec, M. (2000). The chemical reactivity indices in density functional theory, poster contribution at *The 10th Romanian Conference of Physical Chemistry*, Iasi-Romania, September 26-29, Book of Abstracts, **S1**: Po28.

62. Putz, M. V.; Chiriac, A.; Mracec, M. (2001). The chemical field as a basic descriptor for the chemical reactivity controlled by lasers, poster contribution at *Romanian Academy Timisoara Days*, Physical-Chemistry Section, Timisoara-Romania, May 24-25, published in *Annals of West University of Timisoara, Series of Chemistry*, **10**: 943-944.

63. Putz, M. V.; Chiriac, A.; Mracec, M. (2001). The integrated electronegativity as a measure of the basic chemical reactivity exchange, poster contribution at *Romanian Academy Timisoara Days*, Physical-Chemistry Section, Timisoara-Romania, May 24-25, published in *Annals of West University of Timisoara, Series of Chemistry*, **10**: 944-945.

64. Pauling, L. (1932). The nature of the chemical bond IV. The energy of single bonds and the relative electronegativity of atoms, *J. Am. Chem. Soc.*, **54**: 3570-3582.

65. Mulliken, R. S. (1934). A new electroaffinity scale: together with data on valence states and an ionization potential and electron affinities, *J. Chem. Phys.*, **2**: 782-793.

66. Gordy, W. (1946). A new method of determining electronegativity from other atomic properties, *Phys. Rev.*, **69**: 604-607.

67. Iczkowski, R. P.; Margrave, J. L. (1961). Electronegativity, *J. Am. Chem. Soc.*, **83**: 3547-3551.

68. Klopman, G. (1965). Electronegativity, *J. Chem. Phys.*, **43**: S124-S129.

69. Hinze, J.; Jaffe, H. H. (1962). Electronegativity. I. Orbital electronegativity of neutral atoms, *J. Am. Chem. Soc.*, **84**: 540-546.

70. Hinze, J.; Jaffe, H. H. (1963). Electronegativity. IV. Orbital electronegativities of neutral atoms of the period three A and four A and of positive ions of periods one and two, *J. Phys. Chem.*, **67**: 1501-1506.

71. Alonso, J. A.; Grifalco, L. A. (1979). Electronegativity scale for metals, *Phys. Rev. B*, **19**: 3889-3895.

72. Parr, R. G.; Donnelly, R. A.; Levy, M.; Palke, W. E. (1978). Electronegativity: the density functional viewpoint, *J. Chem. Phys.*, **68**: 3801-3807.

73. Bartolotti, L. J.; Gadre, S. R.; Parr, R. G. (1980). Electronegativities of the elements from the simple Xα theory, *J. Am. Chem. Soc.*, **102**: 2945-2948.

74. Robles, J.; Bartolotti, L. J. (1984). Electronegativities, electron affinities, ionization potentials, and hardnesses of the elements within spin polarized density functional theory, *J. Am. Chem. Soc.*, **106**: 3723-3727.

75. Bergmann, D.; Hinze, J. (1987). Electronegativity and charge distribution, *Structure and Bonding*, **66**: 145-190.

76. Gázquez, J. L.; Ortiz, E. (1984). Electronegativities and hardness of open shell atoms, *J. Chem. Phys.*, **81**: 2741-2748.

77. Gázquez, J. L.; Vela, A.; Galvan, M. (1987). Fukui function, electronegativity and hardness in the Kohn-Sham Theory, *Structure and Bonding*, **66**: 79-98.

78. Bartolotti, L. J. (1987). Absolute electronegativities as determined from Kohn-Sham theory, *Structure and Bonding*, **66**: 27-40.

79. Pearson, R. G. (1985). Absolute electronegativity and absolute hardness of Lewis acids and bases, *J. Am. Chem. Soc.*, **107**: 6801-6806.

80. De Proft, F.; Geerlings, P. (1997). Contribution of the shape factor $\sigma(r)$ to atomic and molecular electronegativities, *J. Phys. Chem. A*, **101**: 5344-5346.

81. Parr, R. G.; Pearson, R. G. (1983). Absolute hardness: companion parameter to absolute electronegativity, *J. Am. Chem. Soc.*, **105**: 7512-7516.

82. Murphy, L. R.; Meek, T. L.; Allred, A. L. and Allen, L. C. (2000). Evaluation and test of Pauling's electronegativity scale, *J. Phys. Chem. A*, **104**: 5867-5871.

83. Ray, N. K.; Samuels, L.; Parr, R. G. (1979). Studies of electronegativity equalization, *J. Chem. Phys.*, **70**: 3680-3684.

84. Mortier, W. J.; van Genechten, K.; Gasteiger, J. (1985). Electronegativity equalization: application and parameterization, *J. Am. Chem. Soc.*, **107**: 829-835.

85. Mortier, W. J.; Ghosh, S. K.; Shankar, S. (1986). Electronegativity equalization method for the calculation of atomic charge in molecules, *J. Am. Chem. Soc.*, **108**: 4315-4320.

86. Mortier, W. J. (1987). Electronegativity equalization and its applications, *Structure and Bonding*, **66**: 125-143.

87. Parr, R. G.; Bartolotti, L. J. (1982) On the geometric mean principle of electronegativity equalization, *J. Am. Chem. Soc.*, **104**: 3801-3803.

88. van Genechten, K.; Mortier, W. J.; Geerlings, P. (1987). Intrinsic framework electronegativity: a novel concept in solid state chemistry, *J. Chem. Phys.*, **86**: 5063-5071.

89. Sanderson, R. T. (1988). *Chemical Bonds and Bond Energy*, 2nd edition, Academic Press: New York.

90. Tachibana, A. (1987). Density functional rationale of chemical reaction coordinate, *Int. J. Quantum Chem.*, **21**: 181-190.

91.   Tachibana, A.; Parr, R. G. (1992). On the redistribution of electrons for chemical reaction systems, *Int. J. Quantum. Chem.*, **41**: 527-555.

92.   Tachibana, A.; Nakamura, K.; Sakata, K.; Morisaki, T. (1999). Application of the regional density functional theory: the chemical potential inequality in the HeH^+ system, *Int. J. Quantum Chem.*, **74**: 669-679.

93.   Parr, R. G.; Yang, W. (1984). Density functional approach to the frontier electron theory of chemical reactivity, *J. Am. Chem. Soc.*, **106**: 4049-4050.

94.   Yang, W.; Parr, R. G.; Pucci, R. (1984). Electron density, Kohn-Sham frontier orbitals, and Fukui functions, *J. Chem. Phys.*, **81**: 2862-2863.

95.   Berkowitz, M. (1987). Density functional approach to frontier controlled reactions, *J. Am. Chem. Soc.*, **109**: 4823-4825.

96.   Nalewajski, R. F. (1998). Kohn-Sham description of equilibria and charge transfer in reactive systems, *Int. J. Quantum Chem.*, **69**: 591- 605.

97.   Putz, M. V. (2003). Manuscript in preparation.

98.   Klopman, G. (1968). Chemical Reactivity and the concept of charge and frontier-controlled reactions, *J. Am. Chem. Soc.*, **90**: 223-234.

99.   Chandrakumar, K. R. S.; Pal, S. (2002). A systematic study on the reactivity of Lewis acid-base complexes through the local hard-soft acid-base principle, *J. Phys. Chem. A*, **106**: 11775-11781.

100. Chattaraj, P. K.; Maiti, B. (2003). HSAB principle applied to the time evolution of chemical reactions, *J. Am. Chem. Soc.*, **125**: 2705-2710.

101. Chattaraj, P.; Parr, R. G. (1993). Density functional theory of chemical hardness, *Structure and Bonding*, **80**: 11-25.

102. Ayers, P. W.; Parr, R. G. (2000). Variational principles for describing chemical reactions: the Fukui function and chemical hardness revisited, *J. Am. Chem. Soc.*, **122**: 2010-2018.

103. Chattaraj, P. K.; Liu, G. H.; Parr, R. G. (1995). The maximum hardness principle in the Gyftpoulos-Hatsopoulos three-level model for an atomic or molecular species and its positive and negative ions, *Chem. Phys. Letters*, **237**: 171-176.

104. Mineva, T.; Sicilia, E.; Russo, N. (1998). Density functional approach to hardness evaluation and its use in the study of the maximum hardness principle, *J. Am. Chem. Soc.*, **120**:9053-9058.

105. Ayers, P. W.; Parr, R. G. (2001). Variational principles for describing chemical reactions: reactivity indices based on the external potential, *J. Am. Chem. Soc.*, **123**: 2007-2017.

106. De Luca, G.; Sicilia, E.; Russo, N.; Mineva, T. (2002). On the hardness evaluation in solvent for neutral and charged system, *J. Am. Chem. Soc.*, **124**: 1494-1499.

107. Pearson, R. G. (1997). *Chemical Hardness*, Wiley-VCH: Weinheim.

108. Deb, B. M. (1973). The force concept in chemistry, *Rev. Mod. Phys.*, **45**: 22-43.

109. Bamzai, A. S.; Deb, B. M. (1981). The role of single-particle density in Chemistry, *Rev. Mod. Phys.*, **53**: 95-126.

110. Parr, R. G.; Bartolotti, L. J. (1983). Some remarks on the density functional theory of few-electron systems, *J. Phys. Chem.*, **87**: 2810-2815.

111. Berkowitz, M.; Ghosh, S. K.; Parr, R. G. (1985). On the concept of local hardness in chemistry, *J. Am. Chem. Soc.*, **107**: 6811-6814.

112. Berkowitz, M.; Parr, R. G. (1988). Molecular hardness and softness, local hardness and softness, hardness and softness kernels, and relations among these quantities, *J. Chem. Phys.*, **88**: 2554-2557.

113. Galvan, M.; Vela, A.; Gázquez, J. L. (1988). Chemical reactivity in spin-polarized density functional theory, *J. Phys. Chem.*, **92**: 6470-6474.

114. Vela, A.; Gázquez, J. L. (1988). Extended Hückel parameters from density functional theory, *J. Phys. Chem.*, **92**: 5688-5693.

115. Komorowski, L. (1987). Empirical evaluation of chemical hardness, *Chem. Phys. Letters*, **134**: 536-540.

116. Komorowski, L. (1987). Electronegativity and hardness in chemical approximation, *Chem. Phys.*, **55**: 114-130.

117. Parr, R. G.; Gázquez, J. L. (1993). Hardness functional, *J. Phys. Chem.*, **97**: 3939-3940.

118. Baekelandt, B. G.; Cedillo, A.; Parr, R. G. (1995). Reactivity indices and fluctuations formulas in density functional theory: isomorphic ensembles and a new measure of local hardness, *J. Chem. Phys.*, **103**: 8548-8556.

119. Senet, P. (1996). Nonlinear electronic responses, Fukui functions and hardnesses as functionals of the ground-state electronic density, *J. Chem. Phys.*, **105**: 6471-6489.

120. De Proft, F.; Liu, S. ; Parr, R. G. (1997). Chemical potential, hardness and softness kernel and local hardness in the isomorphic ensemble of density functional theory, *J. Chem. Phys.*, **107**: 3000-3006.

121. Gázquez, J. L; Galván, M; Vela, A. (1990). Chemical reactivity in density functional theory: the N-differentiability problem, *J. Mol. Struct. (Theochem)*, **210**: 29-38.

122. March, N. H. (1998). Density functional theory in relation to X-ray and neutron scattering experiments, *Int. J. Quantum. Chem.*, **69**: 551-557.

123. Chattaraj, P. K.; Maiti, B. (1998). Quantum fluid density functional theory of time-dependent processes, *Int. J. Quantum Chem.*, **69**: 279-291.

124. Mineva, T.; Neshev, N.; Russo, N.; Sicilia, E.; Toscano, M. (1999). Density functional orbital reactivity indices. Fundamentals and applications, *Adv. Quantum Chem.*, **33**: 273-293.

125. Sentilkumar, K.; Ramaswamy, M.; Kolandaivel, P. (2001). Studies of chemical hardness and Fukui function using the exact solution of the density functional theory, *Int. J. Quantum Chem.*, **81**: 4-10.

126. Kolandaivel, P.; Mahalingam, T.; Sugandhi, K. (2002). Polarizability and chemical hardness: a combined study of wave function and density functional theory approach, *Int. J. Quantum Chem.*, **86**: 368-375.

127. Pérez, P.; Andrés, J.; Safont, V. S.; Tapia, O.; Contreras, R. (2002). Spin-philicity and spin-donicity as auxiliary concepts to quantify spin-catalysis phenomena, *J. Phys. Chem. A*, **106**: 5353-5357.

128. Feynman, R. P. (1939). Forces in molecules, *Phys. Rev.*, **56**: 340-343.

129. Garza, J.; Robles, J. (1993). Density functional theory softness kernel, *Phys. Rev. A*, **47**: 2680-2685.

130. Feynman, R. P. (1948). Space-time approach to non-relativistic quantum mechanics, *Rev. Mod. Phys.*, **20**: 367-387.

131. Feynman, R. P. (1972). *Statistical Mechanics*, Addison-Wesley: Redwood City.

132. Giachetti, R.; Tognetti, V. (1985). Variational approach to quantum statistical mechanics of nonlinear systems with application to Sine-Gordon chains, *Phys. Rev. Lett.*, **55**: 912-915; idem (1986). Quantum corrections to the thermodynamics of nonlinear systems, *Phys. Rev. B*, **33**: 7647-7658.

133. Janke, W.; Pelster, A.; Bachmann, M.; Schmidt H.-J., Eds. (2001). *Fluctuating Paths and Fields*, World Scientific: Singapore.

134. Landman, U.; Scharf, D.; Jortner, J. (1985). Electron localization in alkali-halide clusters, *Phys. Rev. Lett.*, **54**: 1860-1863.

135. Tromp, J. W.; Miller, W. H. (1986). New approach to quantum mechanical transition-state theory, *J. Phys. Chem.*, **90**: 3482-3485.

136. Ceperley, D. M.; Jacucci, G. (1987). Calculation of exchange frequencies in bcc ^{3}He with the path-integral Monte Carlo method, *Phys. Rev. Lett.*, **58**: 1648-1651.

137. Gillan, M. J. (1987). Quantum simulation of hydrogen in metals, *Phys. Rev. Lett.*, **58**: 563-566.

138. Schnitker, J.; Rossky, P. J. (1987). Quantum simulation study of the hydrated electron, *J. Chem. Phys.*, **86**: 3471-3485.

139. Parinello, M.; Rahman, A. (1984). Study of an *F* center in molten KCl, *J. Chem. Phys.*, **80**: 860-867.

140. Coker, D. F.; Berne, B. J.; Thirumalai, D. (1987). Path integral Monte Carlo studies of the behavior of excess electrons in simple fluids, *J. Chem. Phys.*, **86**: 5689-5702.

141. Voth, G. A.; Chandler, D.; Miller, W. H. (1989). Rigorous formulation of quantum transition state theory and its dynamical corrections, *J. Chem. Phys.*, **91**: 7749-7760.

142. Liu, Z.; Berne, B. J. (1993). Electron solvation in methane and ethane, *J. Chem. Phys.*, **99**: 9054-9069.

143. Topper, R. Q.; Truhlar, D. G. (1992). Quantum free-energy calculations: Optimized Fourier path-integral Monte Carlo computation of coupled vibrational partition functions, *J. Chem. Phys.*, **97**: 3647-3667.

144. Topper, R. Q.; Tawa, G. J.; Truhlar, D. G. (1992). Quantum free-energy calculations: A three-dimensional test case, *J. Chem. Phys.*, **97**: 3668-3673; (2000); ibidem., **113**: 3930 (E).

145. Topper, R. Q.; Zhang, Q; Liu, Y. –P.; Truhlar, D. G. (1993). Quantum steam tables. Free energy calculations for H_2O, D_2O, H_2S, and H_2Se by adaptively optimized Monte Carlo Fourier path integrals, *J. Chem. Phys.*, **98**: 4991-5005.

146. Topaler, M.; Makri, N. (1994). Quantum rates for a double well coupled to a dissipative bath: Accurate path integral results and comparison with approximate theories, *J. Chem. Phys.*, **101**: 7500-7519.

147. (a) Makri, N.; Makarov, D. E. (1995). Tensor propagator for iterative quantum time evolution of reduced density matrices. I. Theory, *J. Chem. Phys.*, **102**: 4600-4610; (b) Shao J.; Makri, N. (2002). Iterative path integral formulation of equilibrium correlation functions for quantum dissipative systems, *J. Chem. Phys.*, **116**: 507-514.

148. Makri, N. (1998). Quantum dissipative dynamics: a numerically exact methodology, *J. Phys. Chem. A*, **102**: 4414-4427.

149. Mielke, S. L.; Srinivasan, J.; Truhlar, D. G. (2000). Extrapolation and perturbation schemes for accelerating the convergence of quantum mechanical free energy calculations via the Fourier path-integral Monte Carlo method, *J. Chem. Phys.*, **112**: 8758-8764.

150. Mielke, S. L.; Truhlar, D. G. (2001). A new Fourier path integral method, a more general scheme for extrapolation, and comparison of eight path integral methods for the quantum mechanical calculation of free energies, *J. Chem. Phys.*, **114**: 621-630.

151. Fleming, G. R.; Hänggi P., Eds. (1993). *Activated barrier crossing*, World Scientific: Singapore.

152. Levy, M. (1982). Electron densities in search of Hamiltonians, *Phys. Rev. A*, **26**:1200-1208.

153. Teller, E. (1962). On the stability of molecules in the Thomas-Fermi theory, *Rev. Mod. Phys.*, **34**: 627-631.

154. Bartolotti, L. J.; Acharya, P. K. (1982). On the functional derivative of kinetic energy density functional, *J. Chem. Phys.*, **77**: 4576-4585.

155. Bartolotti, L. J. (1982). A new gradient expansion of the exchange energy to be used in density functional calculations on atoms, *J. Chem. Phys.*, **76**: 6057-6059.

156. Deb, B. M.; Chattaraj, P. K. (1989). Density functional and hydrodynamical approach to ion-atom collision through a new generalized nonlinear Schrödinger equation, *Phys. Rev. A*, **39**: 1696-1713.

157. Wang, Y;. Parr, R. G. (1993). Construction of the exact Kohn-Sham orbitals from a given electron density, *Phys. Rev. A*, **47**: R1591-R1593.

158. Zhao, Q.; Parr, R. G. (1992). Quantities $T_s[n]$ and $T_c[n]$ in density functional theory, *Phys. Rev. A*, **46**: 2337-2343.

159. Zhao, Q.; Morrison, R. C.; Parr, R. G. (1994). From electron densities to Kohn-Sham kinetic energies, orbital energies, exchange-correlation potentials, and exchange-correlation energies, *Phys. Rev. A*, **50**: 2138-2142.

160. Rychlewski, J.; Parr, R. G. (1986). The atom in a molecule: a wave function approach, *J. Chem. Phys.*, **84**: 1696-1704.

161. Li, L.; Parr, R. G. (1986). The atom in a molecule: a density matrix approach, *J. Chem. Phys.*, **84**: 1704-1712.

162. Politzer, P.; Lane, P.; Concha, M. C. (2002). Atomic and molecular energies in terms of electrostatic potentials at nuclei, *Int. J. Quantum Chem.*, **90**: 459-463.

163. Wigner, E. (1932). On the quantum correction for thermodynamic equilibrium, *Phys. Rev.*, **40**: 749-759.

164. (a) Szasz, L. (1985). *Pseudopotential Theory of Atoms and Molecules*, John Wiley & Sons: New York; (b) ***Tables of pseudopotential data: http://indy2.theochem.uni-stuttgart.de; (c) Putz, M. V.; Russo, N. (2003). Electronegativity Scales from Chemical Action Concept and Path Integral Formulations, Research Report-Centre of Excellence, University of Calabria, e-print: http://www.hpcc.unical.it/uploads/report3.2.03.doc.

165. (a) Sen, K. D., Ed. (1987). *Electronegativity*, Structure and Bonding-Series, **66**, Springer Verlag: Berlin; (b) ibid., (1993). *Chemical Hardness*, **80**.

166. *** Tables of physical constants: http://physics.nist.gov/cuu/Constants/index.html

167. Silvi, B.; Savin, A. (1994). Classification of chemical bonds based on topological analysis of electron localization functions, *Nature*, **371**: 683-686.

168. Putz, M. V.; Russo, N.; Sicilia, E. (2003). Atomic radii scale and related size properties from density functional electronegativity formulation, *J. Phys. Chem. A*, **107**: 5461-5465.

169. Nalewajski, R. F.; Parr, R. G. (2001). Information theory thermodynamics of molecules and their Hirshfeld fragments, *J. Phys. Chem. A*, **105**: 7391-7400.

SELECTIVE INDEX

Enjoy, cause you don't know from where came

Drink the wine, cause you don't know where to go...

Omar Khayyam – The Quatrains, 28

CPSIA information can be obtained
at www.ICGtesting.com
Printed in the USA
FSOW02n2326221216
28777FS

9 781581 121841